T0300053

Basics of Ramsey Theory

Basics of Ramsey Theory serves as a gentle introduction to Ramsey theory for students interested in becoming familiar with a dynamic segment of contemporary mathematics that combines, among others, ideas from number theory and combinatorics. The core of the book consists of discussions and proofs of the results now universally known as Ramsey's theorem, van der Waerden's theorem, Schur's theorem, Rado's theorem, the Hales–Jewett theorem, and the Happy End Problem of Erdős and Szekeres. The aim is to present these in a manner that will be challenging but enjoyable, and broadly accessible to anyone with a genuine interest in mathematics.

Features

- Suitable for any undergraduate student who has successfully completed the standard calculus sequence of courses and a standard first (or second) year linear algebra course.
- Filled with visual proofs of fundamental theorems.
- Contains numerous exercises (with their solutions) accessible to undergraduate students.
- Serves as both a textbook or as a supplementary text in an elective course in combinatorics and is aimed at a diverse group of students interested in mathematics.

Dr. Veselin Jungić is a Teaching Professor at the Department of Mathematics, Simon Fraser University, Burnaby, British Columbia, Canada.

Dr. Jungić is a 3M National Teaching Fellow and a Fellow of the Canadian Mathematical Society. He is a recipient of several teaching awards. Veselin is one of only a few Canadian mathematicians who has been awarded both the Canadian Mathematical Society Pouliot Award (2020) and the Canadian Mathematical Society Teaching Award (2012).

Dr. Jungić's publications range from education-related opinion pieces to articles based on his teaching practices to Ramsey theory research and outreach papers.

One of Dr. Jungić's accomplishments is the creation of the Math Catcher Outreach Program. Since the early 2010s, the Program has visited hundreds of classrooms, from kindergarten to grade 12, and created learning resources in multiple Indigenous languages. As an invited speaker, Veselin delivered several dozens of the Math Catcher-related workshops and lectures to teachers, academics, and public at the local, national, and international levels.

Basics of Ramsey Theory

Veselin Jungić
Simon Fraser University, Canada

CRC Press
Taylor & Francis Group
Boca Raton London New York

CRC Press is an imprint of the
Taylor & Francis Group, an **informa** business

First edition published 2023
by CRC Press
6000 Broken Sound Parkway NW, Suite 300, Boca Raton, FL 33487-2742

and by CRC Press
4 Park Square, Milton Park, Abingdon, Oxon, OX14 4RN

CRC Press is an imprint of Taylor & Francis Group, LLC

ISBN: 978-1-032-26037-2 (hbk)
ISBN: 978-1-032-26066-2 (pbk)
ISBN: 978-1-003-28637-0 (ebk)

DOI: 10.1201/9781003286370

Typeset in TeXGyreTermes
by KnowledgeWorks Global Ltd.

Publisher's note: This book has been prepared from camera-ready copy provided by the authors

To my sons, my best teachers.

To my son, my best teacher.

Contents

Foreword

The question "What is Ramsey theory about anyway?" is the subject of the first chapter of this unusual and strikingly original book, which is an outcome of an undergraduate course in Ramsey Theory which Veselin Jungić taught for several years at Simon Fraser University.

People variously see the beginning of Ramsey Theory in Hilbert's "Cube Lemma" of 1892, or Schur's Theorem of 1916, or the Baudet–Schur–van der Waerden theorem of 1927, or, indeed, Ramsey's paper of 1930. (There's a long discussion of this in [109].) In any case, the name "Ramsey Theory" became broadly applied after the publication of (the first editions of) *Ramsey Theory* by Ron Graham, Bruce Rothschild, and Joel Spencer [46], and *Rudiments of Ramsey Theory*, by Ron Graham [49], both in 1980.

As an undergraduate in the late 1950s, I myself first encountered van der Waerden's theorem in A. Y. Khinchin's delightful 1945 book *Three Pearls of Number Theory* [65], and to me, one of the outstanding features of Veselin's book is his very clear and detailed elementary proof of van der Waerden's theorem on arithmetic progressions, which follows van der Waerden's original proof and includes wonderful pictures and diagrams (and terminology which I believe was introduced by Imre B. Leader in about 2000). In A. Y. Khinchin's book is the warning "Do not, however, confuse elementary with simple; as you will see, the solutions of all three problems are not very simple, and it will require not a little effort on your part to understand them well and assimilate them." The present text goes a long way, in my opinion, to change "not a little" to "not too terribly much effort." For a beginning student, this is a big change!

Excerpts from some of the student projects in Veselin's course, carried out in small groups of three or four students, appear in the text. As an aside, I may mention that Veselin has received a number of awards for excellence in university teaching, and that he organized the Math Catcher Outreach Program at Simon Fraser. As part of this program, Veselin produced a series of short animated videos, available in English, French, Spanish, and 11 Indigenous languages, aimed at making mathematics attractive to a wide range of audiences.

The carefully done (and very helpful) diagrams and illustrations, the numerous historical remarks, the occasional stories involving the personalities of some of the main characters, the excellent exercises and solutions, make this book a unique and rewarding introduction to an appealing area of mathematics.

Thomas C. Brown
Professor Emeritus
Simon Fraser University

Preface

The purpose of this book is to serve as a gentle introduction to Ramsey theory for those interested in becoming familiar with this dynamic segment of contemporary mathematics that combines, among others, ideas from number theory and combinatorics.

This book is intended to be accessible to undergraduate students, mathematicians, and educators who are curious to learn more about Ramsey theory.

I had three major motivations for writing it:

1. Mathematical: The fundamental Ramsey theory results, like Ramsey's theorem or van der Waerden's theorem, for example, are self-contained.

2. Educational: The story about the beginnings and the development of Ramsey theory and people involved in those processes, like Frank Ramsey and Paul Erdős, for instance, supports the view of mathematics as a living organism and a deeply human endeavour.

3. Personal: I was privileged to witness how Ramsey theory served as a portal into the world of mathematical research to some extraordinarily talented young mathematicians. My hope is that this book might open this portal to some of its readers.

The only real prerequisites to fully grasp the material presented in the book, to paraphrase Fikret Vajzović, a Bosnian mathematician, 1928–2017, are knowing how to read and write and possessing a certain level of mathematical maturity. In other words, my presentation of the material is based on an assumption of the reader's high intelligence and minimum background.

Any undergraduate student who has successfully completed the standard calculus sequence of courses and a standard entry level linear algebra course and has a genuine interest in learning mathematics should be able to master the main ideas presented here.

The core of this book consists of discussions and proofs of the results now universally known as Ramsey's theorem, van der Waerden's theorem, Schur's theorem, Rado's theorem, the Hales–Jewett theorem, and the *Happy End Problem* of Erdős and Szekeres. These are listed in the classic book

Ramsey Theory by Ron Graham, Bruce Rothschild, and Joel Spencer [46] as being at the very heart of Ramsey theory.

It should be mentioned that, except for Schur's theorem, these groundbreaking theorems were established by young people. At the time of publication of their result, Frank Ramsey was to be 27 years old, Bertel van der Waerden was 24, Richard Rado was 27, Alfred Hales was 25, Robert Jewett was 26, Paul Erdős was 22, and George Szekeres was 24. Issai Schur was 41.

I also present brief biographical sketches of Frank Ramsey, Paul Erdős, Bertel van der Waerden, Issai Schur, and Richard Rado.

In addition, the book contains exercises, together with solutions, that may help the reader more completely grasp some of the concepts discussed in the text. The vast majority of those exercises were used as homework problems in the undergraduate course that I taught between 2014 and 2021 at Simon Fraser University. For completeness, the exercises also include guided proofs of some of the important relevant results, such as Folkman's theorem or the Gallai–Witt theorem.

Since this text is based on the class lecture notes for an undergraduate course, the reader will occasionally need the help of a *more knowledgeable other*: a peer, another book or article, or Google.

My wish is to give to the reader both a challenging and enjoyable experience in learning some of the basic facts about Ramsey theory, a relatively new mathematical field. In doing so, I will extensively use drawings to illustrate ideas presented in the various claims and/or their proofs.

This approach is inspired by an old conversation with Jonathan Borwein, a Canadian mathematician, 1951–2016, as depicted in the cartoon to the right. Borwein himself was greatly influenced by George Pólya, a Hungarian mathematician, 1887–1985, and Pólya's seminal book *Mathematical discovery: On understanding, learning, and teaching problem solving* [88].

"Sometimes it is easier to see than to say."

Here is a Pólya quote that Borwein frequently used, see for example [11], when talking about visualization as a tool in learning and doing mathematics:

> A mathematical deduction appears to Descartes as a chain of conclusions, a sequence of successive steps. What is needed for

the validity of deduction is intuitive insight at each step which shows that the conclusion attained by that step evidently flows and necessarily follows from formerly acquired knowledge (acquired directly by intuition or indirectly by previous steps). I think that in teaching [novice learners] we should emphasize intuitive insight more than, and long before, deductive reasoning [88].

Similarly, van der Waerden claimed that "whatever one makes explicit and draws is much easier to remember and to reproduce than mere thoughts" [123].

I follow Pólya and van der Waerden's advice and use illustrations with the aim to emphasize intuitive insight and help the reader perceive the presented concept.

I would like to acknowledge the contributions of my collaborators whose art is included in the book: Bethani L'Heureux (pages 7 and 84); Listiarini Listiarini (pages 5, 46, 50 and 56); Kyra Pukanich (pages 4, 10, 18, 46, 48, 66, 68, 71, 76, 84, and 93); and Simon Roy (pages xiv and 15).

No project such as this can be free from errors or incompleteness. I would be grateful to anyone who points out any typos, errors, or provides any other suggestions on how to improve this manuscript.

Veselin Jungić
Department of Mathematics
Simon Fraser University

Symbols

$\emptyset$	empty set
$\{a, b, c\}$	set containing a, b, and c
$\{x : F(x)\}$	set of all x such that $F(x)$
$x \in A$	x belongs to A
$|A|$	cardinality of set A
$\mathbb{N}, \mathbb{Z}, \mathbb{Q}, \mathbb{R}$	sets of natural numbers, integers, rationals, and reals
$A \cup B$ and $A \cap B$	union and intersection of sets A and B
$A \backslash B$	difference between sets A and B
$A \subset B$	A is a subset of B
$[a, b]$	natural numbers between a and b, including a and b
$f : A \rightarrow B$	function f from A to B
$\lfloor x \rfloor$ and $\lceil x \rceil$	floor and ceiling functions
$\binom{m}{n}$	m choose n
$n!$	n factorial
$a \equiv b (\mathrm{mod}\ n)$	a congruent b modulo n
K_n	complete graph on n vertices
$R(r; m_1, \ldots, m_k)$	Ramsey number
$W(l; k)$	van der Waerden number
$W(l; k_1, \ldots, k_l)$	mixed van der Waerden number
$s(r)$	Schur number
$HJ(k; m)$	Hales–Jewett number
$ES(n)$	Erdős–Szekeres number

CHAPTER 1

Introduction: Pioneers and Trailblazers

RAMSEY THEORY has a well-established place in the mosaic of contemporary mathematics. The Mathematics Subject Classification (MSC) code for Ramsey theory is 05D10. This means that Ramsey theory is classified under Combinatorics, code 05-xx, and more specifically under Extremal Combinatorics, code 05Dxx.

But what is Ramsey theory about? Who was Ramsey? How did Ramsey theory begin and when?

In this chapter we offer several possible descriptions of Ramsey theory. Through a series of examples, we introduce the reader to one of the main goals of Ramsey theory: exploring if a certain pattern is resistant to finite partitions of a given set.

We briefly talk about Paul Erdős and Frank Ramsey, two remarkable people whose mathematical vision and curiosity were essential for establishing Ramsey theory as a captivating field of mathematical exploration in its own right. In addition, we introduce the reader to a few historical facts about the beginnings of Ramsey theory.

1.1   COMPLETE DISORDER IS IMPOSSIBLE

"Complete disorder is impossible."

Theodore S. Motzkin, an Israeli–American
mathematician

1908–1970

DOI: 10.1201/9781003286370-1

What is Ramsey theory?

We start with several possible descriptions of Ramsey theory:

- Ramsey theory is the *mathematics of colouring*. – Alexander Soifer, a Russian-born American mathematician

- Ramsey theory is the study of the preservation of properties under set partitions. – Bruce Landman and Aaron Robertson, American mathematicians

- The fundamental kind of question Ramsey theory asks is: can one always find order in chaos? If so, how much? Just how large a slice of chaos do we need to be sure to find a particular amount of order in it? – Imre Leader, a British mathematician

- If mathematics is a science of patterns, then Ramsey theory is a science of the stubbornness of patterns. – Veselin Jungić, a Canadian mathematician

What is Ramsey theory about?

The following examples should give the reader a taste of a common Ramsey theory scenario:

Step 1: Consider a certain mathematical environment, like an algebraic structure or a graph or the Euclidean plane, for example, and fix a set of objects in it. This fixed set of objects, our *pattern*, may be, for example, a set of all solutions of a given equation or a set of subgraphs of the graph of interest or a family of all sets of points meeting a required condition.

Step 2: Colour the set of all elements in the environment; i.e. partition it into non-overlapping cells.

Step 3: Establish if there is a monochromatic object from the fixed set; i.e. check if there is an object from the fixed set that is composed of elements belonging to the same partition cell.

Example 1.1.1. If the natural numbers are finitely coloured, i.e. the set of natural numbers is partitioned into a finite number of cells, must there exist x, y (with x and y not both equal to 2) with $x + y$ and xy monochromatic, i.e. $x + y$ and xy belong to the same partition cell?

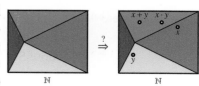

In this example, the *mathematical environment* is the set of all natural numbers, i.e. the set $\mathbb{N} = \{1, 2, 3, \ldots\}$, together with the usual addition and multiplication of natural numbers. The *fixed set of objects* is the set of all two-sets $\{x + y, xy\}$, where x and y are natural numbers not both equal to 2.

The problem was posed by Neil Hindman, an American mathematician, in the late 1970s and affirmatively resolved in 2017 by Joel Moreira, a Portuguese mathematician [82].

What makes this problem *a typical Ramsey theory problem* is the following:

- The topic: the problem is to determine the relationship between the set of all finite partitions of natural numbers and a certain pattern.

- The fact that any numerically literate person can understand the problem.

- It is a difficult problem.

In the next three examples, the mathematical environment will be the *semigroup* $(\mathbb{N}, +)$, i.e. the set of natural numbers, together with the usual rules of addition. The fixed set of objects will be the set of all solutions of a given equation.

In addition, we use these examples to introduce three classical Ramsey theory theorems that will be discussed in detail later in the book.

Example 1.1.2 (Schur's Theorem).

For any partition of the positive integers into a finite number of parts, one of the parts contains three integers x, y, and z with $x + y = z$.

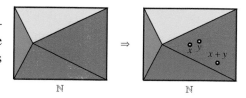

Example 1.1.3 (van der Waerden's Theorem – Special Case).

For any partition of the positive integers into a finite number of parts, one of the parts contains three integers x, y, and z with $x + y = 2z$.

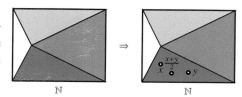

Example 1.1.4 (Rado's Theorem – Special Case).

For any partition of the positive integers into a finite number of parts, one of the parts contains three integers x, y, z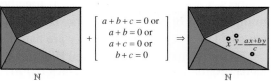

with $ax + by + cz = 0$, $a \neq 0$, $b \neq 0$, $c \neq 0$, if and only if one of the following conditions holds: $a + b + c = 0$ or $a + b = 0$ or $a + c = 0$ or $b + c = 0$.

In the next example, our first encounter with Ramsey's theorem, the mathematical environment will be the so-called complete graph on six vertices, i.e. the graph in which every pair of distinct vertices is connected by a unique edge. The fixed set of objects will be the set of all subgraphs that are complete graphs on three vertices, i.e. the set of all three-sets of edges that form a triangle.

Example 1.1.5 (Ramsey's Theorem – Special Case).

If there are at least six people at dinner, then there are either three mutual acquaintances or three mutual strangers.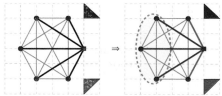

For the purpose of introducing an informal version of the Hales–Jewett theorem, another of the pillars of Ramsey theory, we remind the reader about the game of Tic-Tac-Toe[1]:

Two players who take turns claiming the spaces in a 3×3 grid with the goal to claim a row, a column, or a diagonal.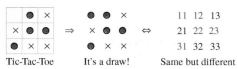

Example 1.1.6 (Hales–Jewett Theorem – Informal).

In large enough dimensions, the game of Tic-Tac-Toe cannot end in a draw.

Robert Jewett, 1937–2022,
and Alfred Hales,
American mathematicians

---

[1]Later we will learn that the setting of the Hales–Jewett theorem excludes one of the diagonals.

1.2 PAUL ERDŐS

"If numbers aren't beautiful, I don't know what is."

Paul Erdős

In this section[2] we provide some brief facts about Erdős's life and work.

1.2.1 Who Was Paul Erdős?

Paul Erdős was a legendary Hungarian-born mathematician who lived as a mathematical pilgrim.

Paul Erdős made contributions to:

Ramsey theory	Graph theory	Discrete mathematics
Classical analysis	Approximation theory	Number theory
Probability theory	Set theory	

Birth and Death. Paul Erdős was born in Budapest, Hungary, on March 26, 1913, and died on September 20, 1996, in Warsaw, Poland.

Anna and Paul Erdős

World in 1913
Srinivasa Ramanujan began a postal correspondence with G.H. Hardy.
The concept of the "isotope" introduced.

First publication of Ludwig Wittgenstein's philosophy of mathematics.
All-purpose zipper patented.

The first four engine aircraft built.
Tesla received a patent for a turbine, known today as the Tesla turbine.
The brand name "Oreo" was registered.
Mahātmā Gandhi arrested for leading Indian miners march in South Africa.
Rabindranath Tagore presented with the Nobel Prize.

World in 1996
The first version of the Java programming language was released.
IBM's Deep Blue wins a game of chess against Gary Kasparov.
The first surface photos of Pluto.

The 34th known Mersenne prime $2^{1257787} - 1$ discovered.
Nintendo 64 goes on sale.
Microsoft releases Internet Explorer 3.0.

Netscape Browser 3.0 is released.
South Africa adopts post-apartheid constitution.
The last federally run Indian residential school closed in Saskatchewan, Canada.

---

[2]This section, as well as Chapter 6, are derived in part from an article published in The Mathematical Intelligencer [58] and used here courtesy of Springer Nature.

Paul's Family. Paul Erdős came from a Jewish family. Paul's mother Anna and father Lajos were teachers of mathematics. Paul's two sisters died of scarlet fever just days before Paul was born. He was raised as an only child.

Erdős in the Words of His Friends and Colleagues. In 1998, in a tribute to Erdős, Lásló Babai, a Hungarian mathematician, wrote:

A mathematical prophet of the jet age, Erdős maintained no permanent home base and was constantly on the move from coast to coast, continent to continent, to visit his ever-growing circle of disciples. He travelled with a small suitcase containing all his earthly belongings. "Property is nuisance," he used to say, paraphrasing the French socialists who thought property was a sin.

Far from being a mathematical robot, Erdős was intensely interested in his human environment. He enjoyed classical music (which he called "noise"), he was well read in history, and he was informed about politics and society. Above all, he cared for the well-being of his friends and colleagues [5].

Thomas C. (Tom) Brown, an American-Canadian mathematician, in a conversation with a group of undergraduate students, said the following in 2020:

Talking with Erdős, or just overhearing him talking with others, was always exciting and even exhilarating. At any big math meeting, if he was talking to one or two people, there would be six or eight people trying to listen in, or trying to ask him a question.

His memory was phenomenal. Here are three examples:

- Once, talking about a result of mine which I thought was new, he simply remarked that he had written a related paper which had appeared in 1938 in such and such a journal. (When I found this paper later, it turned out he had done everything I had done, and more, in a better way.)
- I asked him once in the 80s whether he had ever talked at Reed College. He immediately said something like, "Yes, in April 1955, and I talked about such and such, and had an interesting conversation with so and so."
- He first met my wife Astrid at a math meeting in Israel. (He eventually stayed with us on two occasions for a few days each, during visits to Vancouver.) She had injured her ankle badly in

Cairo the week before. When she next met him, two or three years later, the first thing he said was "How is your ankle now?"

He would often talk while walking. Once at Kitsilano beach [in Vancouver, BC, Canada] he was talking to me and walking, not looking and not noticing where he was going, and he led me right through the middle of a volleyball game. (The players graciously just waited for us to pass by.)

It was hard to keep up with him when trying to solve a problem. He would propose an approach, and quickly develop it, and quickly decide it wouldn't work, drop it, and start on a completely different approach. I would still be trying to grasp his first approach, and he would be half-way to discarding his second approach.

From him I learned that it was dangerous to become attached to a certain approach merely because you had invested a lot of time and energy in it. His ability to develop new ideas and then just drop them to try other new ideas was simply dazzling [108].

"There's this faith in being the scientist and Uncle Paul embodied that faith. He was always searching for truth." – Joel Spencer [70]

Two Saints in St. Gregory of Nyssa Episcopal Church in San Francisco, CA

1.2.2 Erdős's Work: Two Examples

1.2.2.1 Happy End Problem

The tale about the beginnings of the Happy End Problem has been told and retold many times over many years. Together with the problem itself, the tale is an element in mathematical popular culture.

For the purpose of our narration, we go to Budapest, Hungary, on a cold but sunny winter day in the early 1930s. There, in the snow covered Városliget Park, under the statue of Anonymous, we find five friends: Pál "Paul" Erdős, Eszter "Esther" Klein, György "George" Szekeres, Endre "Andre" Makai, and Pál "Paul" Turán.

The friends, all in their late teens and early twenties, are part of a group of young intellectuals that often get together to enjoy their friendship but also to discuss mathematics, politics, art, and the latest developments in science.

We come just in time to hear Esther making the following observation:

> Among any five points in general position in the Euclidean plane, it is always possible to select four points that form the vertices of a convex quadrilateral.

Vocabulary:

Here, *five points in general position in the Euclidean plane* means that no three points lie on the same line. In the figure, five points to the left (blue) are in general position, five points to the right (red) are not.

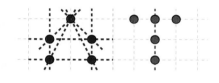

The phrase *a convex quadrilateral* describes a quadrilateral with the property that if two points A and B are inside of the quadrilateral then the whole segment $\overline{AB}$ is inside the quadrilateral. In the figure, the quadrilateral to the left (blue) is convex, the quadrilateral to the right (red) is not.

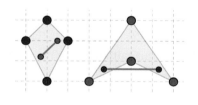

The proof of Esther's observation reduces to distinguish three possible configurations of five points in general position. As the figure shows, each of the three configurations contains vertices of a convex quadrilateral.

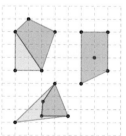

Klein suggested the following more general problem:

> Given any positive integer n, prove that there exists a number $N(n)$ such that among any $N(n)$ points in general position, it is possible to select n points that form the vertices of a convex n-gon.

This is what Szekeres wrote about what happened next:

> I have no clear recollection how the generalization actually came about; in the paper we attributed it to Esther, but she assures me that Paul had much more to do with it. We soon realized that a simple-minded argument would not do and there was a feeling of excitement that a new type of geometric problem emerged from our circle which we were only too eager to solve. For me that it came from Epszi (Paul's name for Esther, short for "epsilon") added a strong incentive to be the first with a solution and after a few weeks I was able to confront Paul with a triumphant 'E. P. open your wise mind.' What I had really found was Ramsey's theorem from which [the generalization] easily followed. Of course, at that time none of us knew about Ramsey [115].

Esther and George married in 1937 in Budapest, Hungary. This is why Erdős named the problem of finding convex n-gons *The Happy End Problem*.

1.2.2.2 *Erdős–Turán Conjecture*

The following statement is commonly known as the Erdős–Turán conjecture:

Conjecture 1.2.1. *If A is a set of positive integers such that $\sum_{n \in A} \frac{1}{n} = \infty$, then A contains arithmetic progressions, $a, a+d, \ldots, a+(k-1)d$, of any given length k.*

In May 1982, Erdős gave an invited lecture entitled *Combinatorial Problems in Geometry* at the 17th New Zealand Mathematics Colloquium. The transcript of Erdős's presentation contains the following motivation for the conjecture as well as Erdős's own assessment of the level of its difficulty:

> An old conjecture in number theory states that there are arbitrarily long arithmetic progressions among the primes. And I think the only way to approach it is this combinatorial conjecture: if you have an infinite sequence of integers $a_1 < a_2 < \ldots$ so that $\sum \frac{1}{a_i} = \infty$ then the sequence contains arbitrarily long arithmetic progressions. This I conjectured more than forty years ago. And since Euler proved that the sum of the reciprocals of the primes diverges, if my conjecture is true it would immediately imply the theorem on primes. So this is a general conjecture: if you have a sequence of integers the sum of whose reciprocals

diverges, for every k you can find $a_{i_1}, \ldots, a_{i_k}$, which form an arithmetic progression of k terms. Now I offer as I said \$3,000 for it, and I said I don't think I will ever have to pay this money and I should leave some money for it in case I leave. The second "leave" means of course leave on the journey where you don't need passports and visas [31].

It turned out that the existence of long arithmetic progressions in the set of primes was established in 2008 by Ben Green, a British mathematician, and Terence Tao, an Australian-born mathematician [51], but that Conjecture 1.2.1 is still unresolved.

In 2020, Thomas Bloom, a British mathematician, and Olof Sisask, a Swedish mathematician, proved the conjecture in the case of three term arithmetic progressions [9].

To claim the award for establishing the truth of the full Conjecture 1.2.1, one should contact Dr. Steve Butler, a professor of mathematics at Iowa State University.

1.3 FRANK PLUMPTON RAMSEY

"Philosophy must be of some use and we must take it seriously; it must clear our thoughts and so our actions."

Frank Plumpton Ramsey

1.3.1 Who Was Frank Ramsey?

Frank Ramsey was a British mathematician, economist, and philosopher. Ramsey made contributions to:

Epistemology	Metaphysics	Philosophy of science
Semantics	Logic	Mathematics
Statistics	Probability	Decision theory
Economics		

Birth and Death. Frank Plumpton Ramsey was born on February 22, 1903, and died on January 19, 1930. Margaret Paul, Ramsey's younger sister, suggests in [87] that the probable cause of his death was a liver illness caused by the Hepatitis B virus that Ramsey contracted while swimming in the River Cam.

World in 1903	World in 1930
$M_{67} = 2^{67} - 1$ is not a prime.	$K_{3,3}$ is not planar.
Bertrand Russell published "The Principles of Mathematics."	Bartel van der Waerden published "Modern Algebra."
Wright Brothers make the first flight.	The first radar detection of planes.
The word "wireless" was included in the Oxford Dictionary.	The world's first radiosonde is launched from Pavlovsk in the Soviet Union.
Jack London published "The Call of the Wild."	Paavo Nurmi ran 6 miles in 29:36.4.
Tesla begins testing at Wardenclyffe tower.	Watty Piper published "The Little Engine That Could."
Nobel for physics awarded to Pierre and Marie Curie.	Nazis gain 107 seats in German election.

Frank's Family. Frank Ramsey came from a distinguished Cambridge family. His parents were Agnes Mary Wilson and Arthur Stanley Ramsey. Frank's father was a mathematician, and the President of Magdalene College. His brother, Arthur Michael Ramsey, became the Archbishop of Canterbury. His sister Bridget was a medical doctor, and his other sister Margaret was a Fellow of Lady Margaret Hall, University of Oxford. In 1925 he married Lattice Baker. They had two daughters, Jane and Sarah.

In 1924, at the age of twenty-one, Ramsey himself got a Fellowship at King's College, Cambridge, having graduated the year before as Cambridge's top mathematics student.

Michael Ramsey about his brother Frank:

He was interested in almost everything. He was immensely widely read in English literature; he was enjoying classics though he was on the verge of plunging into being a mathematical specialist; he was very interested in politics, and well-informed; he had got a political concern and a sort of left-wing caring-for-the-underdog kind of outlook about politics. I was aware that he was far cleverer than I was and knew much more, yet there was such a total lack of uppishness about him that we just conversed in a friendly way and he never made me feel inferior though I was so vastly below par intellectually, and that was the wonderful joy of it [80].

As his sister Margaret indicates in [87], Frank Ramsey was a young renaissance man blessed with an extraordinary intellect, but also an individual who struggled to understand his own personality and emotions. An intellectual who turned everything that he touched to academic gold, but also a person who doubted himself and his work in many ways. A man who was loved and respected by all, but understood by few if any.

Frank Ramsey was a scholar who changed economics and mathematics and continues to provide inspiration to those who seek to better understand the world in which we live. Ramsey's premature death deprived generations of mathematicians, philosophers, and economists from the benefit of his insights.

1.3.2 Ramsey's Work: Two Examples

1.3.2.1 Foundations of Mathematics

Just one year after Ramsey's passing, a collection of his work was published in a book titled *The Foundations of Mathematics and the other Logical Essays*. The book was edited by Richard Braithwaite, an English philosopher, 1900–1990. The collection contained six of Ramsey's already published papers, but about one-half of the book was devoted to Ramsey's previously unpublished essays on various mathematical and philosophical topics.

Braithwaite commented on Ramsey's interest in the foundations of mathematics in the following way:

> Though mathematical teaching was Ramsey's profession, philosophy was his vocation. Reared on the logic of *Principia Mathematica*[3], he was early to see the importance of Dr. Wittgenstein's[4] work and his own published papers were largely based on this. But the previously unprinted essays and notes collected in this volume show him moving towards a kind of pragmatism, and the general treatise on logic upon which at various times he had been engaged was to have treated truth and knowledge as purely natural phenomenon to be explained psychologically without recourse to distinctively logical relations [93].

As an illustration, here is an excerpt from one of Ramsey's posthumously published essays that Braithwaite described as "fragmentary and tentative though they be, [essays] seem to display Ramsey's mind in its highest power."

> I have always said that a belief was knowledge if it was (i) true, (ii) certain, (iii) obtained by a reliable process. But the word 'process' is very unsatisfactory; we can call inference a process, but even then unreliable seems to refer only to a fallacious method not to false premise

---

[3]*Principia Mathematica* is a three-volume landmark work on the foundations of mathematics written by Alfred Whitehead, English mathematician and philosopher, 1861–1947, and Bertrand Russell, a British philosopher, 1872–1970.

[4]Ludwig Wittgenstein, an Austrian philosopher, 1889–1951.

as it is supposed to do. Can we say that a memory is obtained by a reliable process? I think perhaps we can if we mean the causal process connecting what happens with my remembering it. We might then say, a belief obtained by a reliable process must be caused by what are not beliefs in a way or with accompaniments that can be more or less relied on to give true beliefs, and if in this train of causation occur other intermediary beliefs these must all be true ones [93].

1.3.2.2 Ramsey's Theorem

Consider the global population at the present and imagine that you can form all possible groups of, for example, ten people. Next, partition this newly formed set of groups of 10 people into, for example, 100 mutually disjoint cells following any criterion you prefer. Can you be sure that there would be, for example, 1,000 people so that all groups of 10 that contain only individuals from those chosen 1,000 people belong to the same partition cell?

Yes, if there were enough people on the Earth, because according to Ramsey's Theorem:

Theorem 1.3.1 (Ramsey's Theorem [92]). *Given any r, n, and μ we can find an m_0 such that, if $m \geq m_0$ and the r-combinations of any Γ_m are divided in any manner into μ mutually exclusive classes C_i ($i = 1, 2, \ldots, \mu$), then Γ_m must contain a sub-class Δ_n such that all the r-combinations of members of Δ_n belong to the same C_i.*

This is Ramsey's authentic statement of the theorem. Ramsey used Γ_m to denote a set with m elements.

We observe that the mathematical jargon has changed over time. A modern mathematician would likely say, for example, *r-subset* instead of *r-combination*; *mutually disjoint sets* instead of *mutually exclusive classes*; *an n-subset Δ_n* instead of *a sub-class Δ_n*; *r-subsets of Δ_n* instead of *r-combinations of members of Δ_n*; and *elements of Δ_n* instead of *members of Δ_n*.

In our example, $r = 10$, $n = 1,000$, $\mu = 100$, and m represents the size of the global population Γ_m at a certain moment in time. For these values of r, n, and μ, the value of the number m_0 is unknown, but almost certainly the world population will never reach the required m_0.

1.4   RAMSEY THEORY

"Ramsey's theorem was not discovered by Paul Erdős. But perhaps one could say that Ramsey theory was created largely by him."

Ronald L. (Ron) Graham, an American mathematician, 1935–2020, and Jaroslav Nešetřil, a Czech mathematician

Today Ramsey's theorem is one of the cornerstones of Ramsey theory. Some other results that form the very base of Ramsey theory are Hilbert's cube lemma (1892), Schur's theorem (1916), and van der Waerden's theorem (1927).

This clearly contradicts the statement by David Hugh Mellor, a British philosopher, 1938–2020, that "Frank P. Ramsey will be known to readers of the Journal of Graph Theory as the eponymous discoverer of Ramsey numbers and founder of Ramsey theory [79] ."

So, when did Ramsey theory become *Ramsey* theory?

Soifer offers in [109] a detailed account of his own investigation about this question and concludes:

It seems that [Ramsey theory] has been shaping throughout the 1970s, and the central engine of this process was new results and two surveys by Graham and Rothschild. In 1980 the long life of the name was assured when it appeared as the title of the book *Ramsey Theory* by Graham, Rothschild, and Spencer [109, p. 290].

When asked in 2021 by a group of undergraduate students if he and his coauthors Ron Graham and Bruce Rothschild knew that the Ramsey theory book would be so influential, Joel Spencer, an American mathematician, replied:

Well, we hoped. I mean, we didn't know.

The way we looked at it at the time, we saw that there were already some beautiful theorems. There was Ramsey's theorem itself. There was van der Waerden's theorem. There was the Hales–Jewett theorem. And there were some others, it was like a collection of theorems, but it wasn't even yet a collection.

And yet we felt that these were all in the same area. So we called it Ramsey theory and we named the book and it's stuck.

I mean we deliberately named the book Ramsey theory with the idea that we were taking these separate theorems, all in their separate papers, that people had been reading. Our major contribution in writing the book was to say, look, this is a cohesive area. Here are these different theorems, but they have common methodologies and this common theme to them.

We succeeded really beyond my wildest dreams as now that's just common knowledge. But it wasn't common knowledge when we started. So I feel very good about it. And you two are good examples of it, because now you're studying Ramsey theory. Now, the field is there and there are people working on it.

Did I predict that it would be so successful? No, I didn't. I had hope [70].

As of the birthplace of Ramsey theory, in 1983 Spencer claimed: "In my opinion, Ramsey theory was born, after a long and healthy embryonic stage, at the Combinatorial Conference at Balatonfüred, Hungary, 1973 [111]."

Who was the father? In the preface to the first edition of the book *Ramsey Theory*, the authors attribute Paul Erdős as one "who can rightfully be considered the father of modern Ramsey theory."

"The stars may be large, but they cannot think or love; and these are qualities which impress me far more than size does."
- Frank P. Ramsey

CHAPTER 2

Ramsey's Theorem

Ramsey's THEOREM is one of the cornerstones of Ramsey theory. In this chapter, we will discuss several special cases as well as a general case of Ramsey's theorem.

We will mostly work with, what Ramsey called, two-combinations of Γ_m, i.e. with the set of all subsets of Γ_m that contain exactly two elements. In this setting, we think about the set Γ_m as the set of vertices of a complete graph K_m and we represent a two-combination $\{A, B\} \subseteq \Gamma_m$ by the corresponding edge, i.e. by the edge in the graph K_m that is incident with the vertices A and B.

We start with the so-called *pigeonhole principle* for two reasons. Firstly, the pigeonhole principle *is* a special case of Ramsey's theorem. Secondly, the pigeonhole principle is one of the fundamental tools that will be used throughout this book.

This is followed by our search for solutions of three instances of the so-called *dinner party problem*. Our search will lead us to the establishment of several values of *Ramsey numbers*. This will open a window through which we will be able to reach out to two fascinating and self-standing statements, which are special cases of Ramsey's theorem, namely Ramsey's theorem for graphs and a particular case of the infinite Ramsey's theorem.

Finally, we will prove Ramsey's theorem.

2.1 THE PIGEONHOLE PRINCIPLE

> "There are three kinds of mathematicians: Those who know how to count and those who don't."
>
> Anonymous

DOI: 10.1201/9781003286370-2

Theorem 2.1.1 (The Pigeonhole Principle). *Suppose you have k pigeonholes and n pigeons to be placed in them. If n > k then at least one pigeonhole contains at least two pigeons.*

The pigeonhole principle has been attributed to Johann Peter Gustav Lejeune Dirichlet, a German mathematician, 1805–1859.

On its surface, the pigeonhole principle is intuitively clear: if there are 10 pigeons and 9 pigeonholes, then at least one pigeonhole contains at least two pigeons.

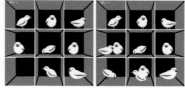

Under the surface, the pigeonhole principle reflects one of the questions that has been a source of debate among logicians and philosophers since the last quarter of the 19th century: "How are natural numbers individuated? That is, what is our most basic way of singling out a natural number for reference in language or in thought [76]?"

One of those *basic ways of singling out a natural number* is the notion of cardinal numbers: two sets are assigned the same cardinal number if there is a one-to-one correspondence between them. In this setting, natural numbers are cardinal numbers of non-empty finite sets, i.e. sets that are not in a one-to-one correspondence with any of their proper subsets.

Let n be the cardinal number of a finite set A and let k be the cardinal number of a finite set B. If A contains a proper subset A' with k as its cardinal number, then we say that n is greater than k and write $n > k$. In other words, $n > k$ means that for any function $f : A \to B$, there are $x, y \in A$, $x \neq y$, and $z \in B$ such that $f(x) = f(y) = z$. Or, in the terms of the pigeonhole principle, at least two pigeons (x and y) belong to the same pigeonhole (z).

Example 2.1.1. Show that among any 5 integers there are two such that their difference is divisible by 4.

Solution. Say that there are four pigeonholes: 0, 1, 2, and 3. We put the integer a in the pigeonhole i, $i \in \{0, 1, 2, 3\}$, if i is the remainder when a is divided by 4. For example, we put the number $27 = 6 \cdot 4 + 3$ in the pigeonhole "3."

Since there are 5 integers, by the pigeonhole principle, at least two of them must go to the same pigeonhole: $a = 4k + i$ and $b = 4n + i$. It follows that $a - b = (4k + i) - (4n + i) = 4(k - n)$ is divisible by 4. □

Example 2.1.2. Consider a chess board with two of the diagonally opposite corners removed. Is it possible to cover the board with dominoes of size exactly two board squares?

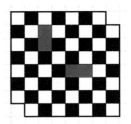

Solution. Observe that there are sixty-two 1×1 squares on this chess board. Hence, thirty-one 2×1 dominos would be needed to cover the board. Also observe that the new board contains thirty-two white squares and that each domino covers one white and one black square.

Consider thirty-one dominos as the pigeonholes and the white squares as the pigeons. Since there are thirty-two white squares, by the pigeonhole principle, at least one domino would have to cover two white squares, which is impossible. □

Example 2.1.3. A grid of twenty-seven points in the plane is given. Each point is coloured red or black. Prove that there exists a monochromatic rectangle, i.e. a rectangle with all four vertices of the same colour.

Solution. Observe that there are eight distinct ways to colour three points with two colours. Also observe that, by the pigeonhole principle, each coloured column contains at least two points of the same colour.

Let the grid be coloured red and black in any of $2^{3 \cdot 9} = 2^{27} = 134,217,728$ ways. Let the nine columns be the pigeonholes. Since there are only eight distinct ways to colour a column and since there are nine columns, by the pigeonhole principle, there must be at least two of the columns coloured in the same way.

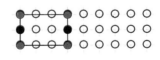

A monochromatic rectangle appears! □

The pigeonhole principle establishes that if there are more pigeons than pigeonholes then at least one pigeonhole must contain at least two pigeons. But, can we say more? For example, can we, knowing the number of pigeons and the number of pigeonholes, be sure if there must a pigeonhole with more than two pigeons?

Before we state the so-called generalized pigeonhole principle, we recall that the floor function, floor : $\mathbb{R} \to \mathbb{Z}$, is defined by $\text{floor}(x) = \lfloor x \rfloor = \max\{m \in \mathbb{Z} : m \leq x\}$.

In other words, for $x \in \mathbb{R}$, $\lfloor x \rfloor$ is the largest integer that is not greater than x. For example, $\lfloor \pi \rfloor = 3$ and $\lfloor -\pi \rfloor = -4$.

Similarly, the ceiling function, ceil : $\mathbb{R} \to \mathbb{Z}$, is defined by $\text{ceil}(x) = \lceil x \rceil = \min\{m \in \mathbb{Z} : x \leq m\}$.

Hence, for $x \in \mathbb{R}$, $\lceil x \rceil$ is the smallest integer that is not smaller than x. For example, $\lceil \pi \rceil = 4$ and $\lceil -\pi \rceil = -3$.

Theorem 2.1.2 (Generalized Pigeonhole Principle). *If n pigeons are sitting in k pigeonholes, where n > k, then there is at least one pigeonhole with at least $\lceil \frac{n}{k} \rceil$ pigeons and at least one pigeonhole containing not more than $\lfloor \frac{n}{k} \rfloor$ pigeons.*

Proof. By definition, $\lceil \frac{n}{k} \rceil$ is the integer with the property that $\frac{n}{k} \leq \lceil \frac{n}{k} \rceil < \frac{n}{k} + 1$.

If none of the k pigeonholes contains $\lceil \frac{n}{k} \rceil$ pigeons, i.e. if the maximum number of the pigeons per pigeonhole is less than or equal to $\lceil \frac{n}{k} \rceil - 1$ then

$$\text{(the number of pigeons)} \leq k \cdot \left(\left\lceil \frac{n}{k} \right\rceil - 1 \right) < k \cdot \left(\left(\frac{n}{k} + 1 \right) - 1 \right) = k \cdot \frac{n}{k} = n,$$

which contradicts the assumption that there were n pigeons.

Similarly, since $\frac{n}{k} - 1 < \lfloor \frac{n}{k} \rfloor \leq \frac{n}{k}$, if each of the k pigeonholes contains more than $\lfloor \frac{n}{k} \rfloor$ pigeons, then

$$\text{(the number of pigeons)} \geq k \cdot \left(\left\lfloor \frac{n}{k} \right\rfloor + 1 \right) > k \cdot \left(\left(\frac{n}{k} - 1 \right) + 1 \right) = k \cdot \frac{n}{k} = n,$$

which again contradicts the assumption that there were n pigeons. □

Example 2.1.4. There are thirteen non-overlapping time periods in a weekday during which classes at a university can be scheduled. If there are 672 different classes that need to be scheduled on each Wednesday, how many different rooms will be needed?

Solution. Here, $n = 672$ classes (pigeons) and $k = 13$ different time slots (pigeonholes). By the generalized pigeonhole principle, there is at least one pigeonhole (time slot) with at least $\left\lceil \frac{n}{k} \right\rceil = \left\lceil \frac{672}{13} \right\rceil = \left\lceil 51 \frac{9}{13} \right\rceil = 52$ pigeons.

Observe that $51 \cdot 13 = 663 < 672 < 52 \cdot 13 = 676$. Hence at least 52 class-rooms are needed to schedule 672 classes on each Wednesday. □

2.2 ACQUAINTANCES AND STRANGERS

"A friend to all is a friend to none."

Aristotle, a Greek philosopher

384 BCE–322 BCE

We start our study of Ramsey's theorem by discussing the following instance of the so-called dinner party problem:

Question 2.2.1 (Dinner Party Problem). Suppose that six people are gathered at a dinner party. Can we be sure that there is a group of three people at the party who are either all mutual acquaintances or all mutual strangers?

Our strategy will be to represent each person at the party as a vertex of K_6, the complete graph on six vertices, see Example 2.2.1. We will colour each edge by one of the two available colours, depending on whether the two people, represented by the vertices incident to the edge, are two acquaintances or two mutual strangers.

Example 2.2.1 (Edge Two-Colouring). Use two colours, red and blue, for example, to colour the edges of K_6. Each edge should be coloured by only one colour.

K_6 – a complete graph on six vertices

Question 2.2.2. Consider the following two questions:

(a) How many distinct edge two-colourings of K_6 are there?

(b) Can you find a monochromatic triangle in your colouring, i.e. can you find three edges coloured by the same colour that form a triangle?

Since there are fifteen edges in K_6 and since for each edge there is a choice between two colours, the answer to the first question is $2^{15} = 32,768$.

The answer to the second question should be a "yes."

BIG Question:

Does any edge two-colouring of K_6 contain a monochromatic triangle?

BIG Answer: Yes, any edge two-colouring of K_6 contains a monochromatic triangle!

Theorem 2.2.1 (Ramsey's Theorem – Special Case). *Any edge two-colouring of K_6 contains a monochromatic K_3.*

Proof. Let an edge two-colouring of K_6 be given.

We fix one of the vertices in K_6 and think about the five edges incident to the fixed vertex as pigeons. We think about the two colours, say blue and red, as the pigeon-holes.

By the (generalized) pigeonhole principle, at least three of these five edges are of the same colour, say blue. Three vertices that are adjacent to the fixed vertex by three blue edges, determine a K_3.

What if there is a blue edge in this K_3?

What if there is no blue edge in this K_3?

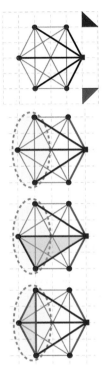

□

Question 2.2.3. Can we replace K_6 in Theorem 2.2.1 by K_5?

Thus the answer to Question 2.2.1 is yes. But if there had been only five people at the dinner party then the answer would have been no.

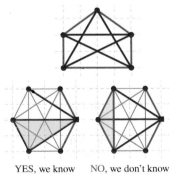

YES, we know   NO, we don't know
each other   each other

What if we consider a different instance of the dinner party problem?

Question 2.2.4. Suppose that ten people are gathered at a dinner party. Can we be sure that there is a group of *four* people at the party who are all mutual acquaintances or a group of *three* people who are all mutual strangers?

And if the answer is a yes, what would happen if we consider a smaller party, with nine or eight people, for example?

We explore the above questions in several steps.

Claim 2.2.1. *Any edge two-colouring (red and blue) of K_{10} contains a red K_4 or a blue K_3.*

Proof. Consider any edge two-colouring (red and blue) of K_{10}.

Observe that, by the pigeonhole principle, each vertex is incident to at least six red edges or to at least four blue edges.

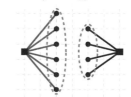

If there is a vertex incident to at least four blue edges then that vertex belongs to a blue K_3 or it is adjacent to four vertices that belong to a red K_4.

If there is a vertex incident to at least six red edges then, by Theorem 2.2.1, that vertex is adjacent to three vertices that belong to a blue K_3 or it belongs to a red K_4.

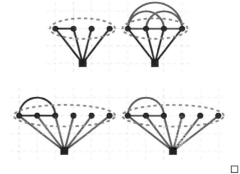

□

Claim 2.2.2. *Any edge two-colouring (red and blue) of K_9 contains a red K_4 or a blue K_3.*

Proof. Consider any edge two-colouring (red and blue) of K_9. Note that each vertex in K_9 is incident to exactly eight edges.

If there is a vertex incident to at least six red edges or to at least four blue edges then we have one of the cases discussed in Claim 2.2.1.

Otherwise, each vertex would be incident to exactly five red edges and exactly three blue edges.

If this is the case then the total number of the blue edges would be:

$$\frac{(\#\text{ of vertices })\cdot(\#\text{ of incident blue edges })}{2} = \frac{9\cdot 3}{2} = 13.5.$$

But, since the number of the blue edges must be an integer, this is not possible.

Hence there must be a vertex that is incident to at least six red edges or to at least four blue edges. □

Question 2.2.5. Does every red-blue edge colouring of K_8 contain a red K_4 or a blue K_3?

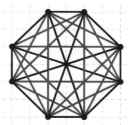

Putting everything together, we obtain another special case of Ramsey's theorem.

Theorem 2.2.2 (Ramsey's Theorem – Special Case). *Any red-blue edge colouring of K_9 contains a red K_4 or a blue K_3.*

Moreover, we have answered Question 2.2.4: there should be at least nine people at the dinner party to make sure that there is a group of four people who are all mutual acquaintances or a group of three people who are all mutual strangers.

Commonly, the answer to Question 2.2.4 is written as $R(4,3) = 9$. In this notation the answer to Question 2.2.1 can be restated as $R(3,3) = 6$.

We refine the statement of Theorem 2.2.2.

Theorem 2.2.3 (Ramsey's Theorem – Special Case). $R(4,3) = R(3,4) = 9$.

We finish this section by searching for $R(4,4)$, i.e. by solving the following instance of the dinner party problem:

Question 2.2.6. What is the minimum number of people at a dinner party so that we can be sure that there is a group of *four* people at the party who are all mutual acquaintances or a group of *four* people who are all mutual strangers?

Theorem 2.2.4 (Ramsey's Theorem – Special Case). $R(4,4) \leq 18$.

Proof. Consider a red-blue colouring of a K_{18}.

Observe that each vertex ■ is incident to at least nine edges of the same colour, say red.

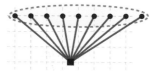

Case 1: There is a red K_3 in the induced K_9.

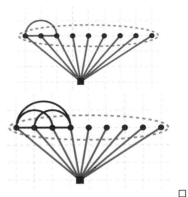

Case 2: There is a blue K_4 in the induced K_9.

□

Moreover:

Theorem 2.2.5 (Ramsey's Theorem – Special Case). $R(4,4) = 18$.

Proof. See Exercise 2.14. □

Therefore, the answer to Question 2.2.6 is: there should be at least eighteen people at the dinner party to make sure that there is a group of four people at the party who are either all mutual acquaintances or all mutual strangers.

2.3 RAMSEY'S THEOREM FOR GRAPHS

"To see things in the seed, that is genius."

Laozi, a Chinese philosopher

6th century BC

As part of the introduction of Ramsey's theorem, in Section 2.2 we considered three instances of the dinner party problem. In the course of our investigation, we established that $R(3,3) = 6$ (Theorem 2.2.1), $R(3,4) = R(4,3) = 9$ (Theorem 2.2.2), and $R(4,4) = 18$ (Theorem 2.2.5).

In this section we take our investigation a step further and prove a more general, but still special, case of Ramsey's theorem.

We start by putting the numbers $R(3,3)$, $R(3,4)$, $R(4,3)$, and $R(4,4)$ into a wider context.

Definition 2.3.1 (Ramsey Number). *The Ramsey number $R(s,t)$ is the minimum number n for which any edge two-colouring of K_n, a complete graph on n vertices, in red and blue, contains a red K_s or a blue K_t.*

For example, $R(s,2) = s$ and $R(2,t) = t$, for any $s,t \in \mathbb{N}\backslash\{1\}$.

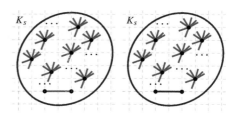

Three BIG Questions:

1. Does the Ramsey number $R(s,t)$ exist for any choice of natural numbers $s \geq 2$ and $t \geq 2$?

2. If $R(s,t)$ exists, is the exact value of $R(s,t)$ known?

3. If $R(s,t)$ exists and if the exact value of $R(s,t)$ is not known, what are the best-known bounds for $R(s,t)$?

The affirmative answer to the first of the big questions above is given by the following version of Ramsey's theorem:

Theorem 2.3.1 (Ramsey's Theorem For Graphs, Two Colours). *For any $s,t \in \mathbb{N}\backslash\{1\}$ the Ramsey number $R(s,t)$ exists and, for $s,t \geq 3$,*

$$R(s,t) \leq R(s-1,t) + R(s,t-1).$$

Before we proceed with the proof of Theorem 2.3.1, we make several observations.

Observation 2.3.1. From $R(2,2) = 2$, $R(3,2) = R(2,3) = 3$, $R(3,3) = 6$, $R(4,2) = R(2,4) = 4$, it follows that, for any $s,t \in \mathbb{N}\backslash\{1\}$ such that $s+t = 4$ or $s+t = 5$ or $s+t = 6$, the Ramsey number $R(s,t)$ exists.

Observation 2.3.2. Since, for any $s \geq 2$, $R(s,2) = R(2,s) = s$, we are interested only in the question if $R(s,t)$ exists for $s,t \geq 3$.

Observation 2.3.3. We recall that $\mathbb{N}$ is a well-ordered set, i.e. that every non-empty set of natural numbers has a least element (in the usual ordering of $\mathbb{N}$). This fact implies that to prove that $R(s,t)$, $s,t \geq 3$, exists it is enough to find *one* natural number M with the property that any two-colouring, say red and blue, of a complete graph K_M contains a monochromatic (red) K_s or a monochromatic (blue) K_t.

Our strategy is to establish, via induction on the sum $s + t$, that any two-colouring of a complete graph K_M, where $M = R(s - 1, t) + R(s, t - 1)$, contains a red K_s or a blue K_t.

This approach, as well as the claim of Theorem 2.3.2 below, comes from the very cradle of Ramsey theory as we know it, the article entitled *A combinatorial problem in geometry*, authored by Paul Erdős and George Szekeres, and published in 1935 [28].

Proof of Theorem 2.3.1. As per Observation 2.3.2 we take $s, t \geq 3$. We use mathematical induction on the sum $s + t$ to prove that $R(s, t)$ exists.

The base case of induction, $s + t = 6$, is established by Observation 2.3.1.

As induction hypothesis we assume that $n \geq 6$ is with the property that, for any $u, v \geq 3$ such that $u + v = n$, the Ramsey number $R(u, v)$ exists.

Let $s, t \geq 3$ be such that $s + t = n + 1$. Since $(s - 1) + t = s + (t - 1) = n$, by the induction hypothesis, $R(s - 1, t)$ and $R(s, t - 1)$ exist. Let $M = R(s - 1, t) + R(s, t - 1)$.

We consider an edge two-colouring of K_M. Observe that each vertex is incident to $M - 1 = R(s - 1, t) + R(s, t - 1) - 1$ edges.

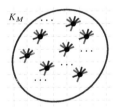

We fix a vertex and count the number of red and the number of blue edges incident to the fixed vertex.

By the pigeonhole principle, there are two possible cases:

There are at least $R(s - 1, t)$ red edges *or* there are at least $R(s, t - 1)$ blue edges incident to the fixed vertex.

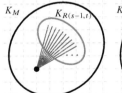
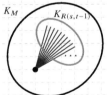

Suppose that there are at least $R(s - 1, t)$ red edges. Recall that, by definition, $K_{R(s-1,t)}$ contains a red K_{s-1} or a blue K_t.

Say that $K_{R(s-1,t)}$ contains a red K_{s-1}. Recall that all edges between the vertices of K_{s-1} and the initially fixed vertex are red. A red K_s appears!

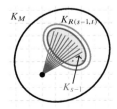

The remaining case is that $K_{R(s-1,t)}$ contains a blue K_t. But, since $K_{R(s-1,t)}$ is a subgraph of K_M, this means that K_M contains a blue K_t.

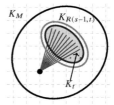

Hence, any red/blue edge two-colouring of K_M contains a red K_s or a blue K_t.

By the principle of mathematical induction, $R(s,t)$ exists for any $r, s \geq 3$ and $R(s,t) \leq R(s-1,t) + R(s,t-1)$. □

The list of at this moment known exact values of $R(s,t)$ is quite short. See Table 2.1.

Table 2.1   Known Ramsey Numbers $R(s,t)$.

s	t	$R(s,t)$	Who and When
3	3	6	
3	4	9	Greenwood and Gleason, 1955
3	5	14	Greenwood and Gleason, 1955
3	6	18	Graver and Yackel, 1968
3	7	23	Kalbfleisch, 1966
3	8	28	McKay and Min, 1992
3	9	36	Grinstead and Roberts, 1982
4	4	18	Greenwood and Gleason, 1955
4	5	25	McKay and Radziszowski, 1995

Even establishing and narrowing the intervals that contain particular Ramsey numbers has been an ongoing challenge for generations of mathematicians and computer scientists. See Table 2.2.

David Conlon, an Irish mathematician, described learning about Ramsey numbers when he was 15 years old:

And on the third or fourth page [of a book by Derek Holton] they mentioned that somehow this was a difficult problem: the Ramsey number for three [Conlon was talking about the so-called diagonal Ramsey numbers $R(s,s)$] is six, the Ramsey number for four is 18, but that Ramsey number for five is just not known. And in general, it's not known. Learning that even good estimates on these things are not known sort of blew my mind at the time. Like how could something this simple, actually not be known [64]?

Table 2.2 More Known Facts.

s	t	$R(s,t)$	Who and When
3	10	[40, 42]	Exoo 1989, Radziszowski and Kreher 1988
3	11	[46, 51]	Radziszowski and Kreher 1988
4	6	[35, 41]	Exoo, McKay and Radziszowski 1995
4	7	[49, 61]	Exoo 1989, Mackey 1994
5	5	[43, 48]	Exoo 1989, McKay and Radziszowski 1995
5	6	[58, 87]	Exoo 1993, Walker 1971
6	6	[102, 165]	Kalbfleisch 1965, Mackey 1994
6	7	[113, 298]	Exoo and Tatarevic 2015, Xu and Zhang 2002

For Conlon this early experience with Ramsey numbers was a sign of things that were coming: as part of his doctoral thesis, Conlon made a break-through in lowering the upper bound for the diagonal Ramsey numbers [17].

But before we use Conlon's own words to describe his contribution to a better understanding of the size of Ramsey numbers, we revisit two classical results.

Theorem 2.3.2 (Ramsey Numbers – An Upper Bound). *For any $s, t \in \mathbb{N} \setminus \{1\}$,* $R(s,t) \le \binom{s+t-2}{t-1}$.

Proof. We use mathematical induction. Recall that, for $s \ge 2$, $\binom{s+2-2}{2-1} = \binom{s}{1} = s = R(s,2)$ and that $\binom{3+3-2}{3-1} = \binom{4}{2} = 6 = R(3,3)$.

Hence, if $s, t \ge 2$ and $s + t \le 6$ then the inequality $R(s,t) \le \binom{s+t-2}{t-1}$ holds.

As the induction hypothesis we assume that $n \ge 6$ has the property that the inequality holds for all u, v such that $u + v = n$.

Let $s, t \in \mathbb{N} \setminus \{1,2\}$ be such that $s + t = n + 1$. Observe that this implies $(s-1) + t = s + (t-1) = n$ and that by the induction hypothesis

$$R(s-1,t) \le \binom{(s-1)+t-2}{t-1} = \binom{s+t-3}{t-1}$$

and

$$R(s,t-1) \le \binom{s+(t-1)-2}{(t-1)-1} = \binom{s+t-3}{t-2}.$$

It follows that

$$R(s,t) \leq R(s-1,t) + R(s,t-1)$$
$$\leq \binom{s+t-3}{t-1} + \binom{s+t-3}{t-2} = \binom{s+t-2}{t-1},$$

which completes the induction step.[1]   □

In 1947, Erdős established a lower bound for the diagonal Ramsey numbers $R(s,s)$ [26]. Before we, for the purpose of this text, expand Erdős's 12-lines long original proof, we make a few observations.

Observation 2.3.4. A complete graph of n vertices has $\binom{n}{2} = \frac{n(n-1)}{2}$ edges. This implies that the number of distinct graphs on a given set of n vertices is $2^{\frac{n(n-1)}{2}}$.

Observation 2.3.5. For $n, k \in \mathbb{N}$, $n > k > 1$, $\binom{n}{k} = \frac{n(n-1)\cdots(n-k+1)}{k!} < \frac{n^k}{k!}$.

Observation 2.3.6. For $k \geq 2$, $2^{k+1} < (2 \cdot 1) \cdot (2 \cdot 2) \cdot (2 \cdot 3) \cdot \ldots (2 \cdot k) < (2k)!$

Theorem 2.3.3 (Diagonal Ramsey Numbers – Lower Bound). *For $s \geq 3$, $R(s,s) > 2^{s/2}$.*

Proof. Note that $R(3,3) = 6 > 2^{\frac{3}{2}}$, $R(4,4) = 18 > 2^{\frac{4}{2}}$, and $R(5,5) \geq 43 > 2^{\frac{5}{2}}$. Let $s \geq 6$ and let $n \in \mathbb{N}$ be such that $s < n \leq 2^{s/2}$.

Let $\mathcal{N}$ be the set of all distinct n-vertex subgraphs of K_n.

For $G \in \mathcal{N}$, let $G^c \in \mathcal{N}$ denote the complement of G, i.e. let G^c be a graph with the property that, for any two vertices a and b, $\{a,b\}$ is an edge in G^c if and only $\{a,b\}$ is not an edge in G. For example, the complement of the complete graph K_n is a graph of n vertices and with no edges.

Let PH $= \{\{G, G^c\} : G \in \mathcal{N}\}$. By Observation 2.3.4, the number of elements of the set PH is $|\text{PH}| = \frac{1}{2} \cdot 2^{\frac{n(n-1)}{2}} = 2^{\frac{n(n-1)}{2}-1}$.

Now, let $\mathcal{N}_s$ be the set of all elements of $\mathcal{N}$ that contain a complete graph of s vertices.

We think about the elements of the set $\mathcal{N}_s$ as being pigeons and the elements of the set PH as being pigeonholes: we put the pigeon $G \in \mathcal{N}_s$ into the pigeonhole $\{G, G^c\} \in \text{PH}$.

[1]We applied the fact that $\binom{m}{k} + \binom{m}{k-1} = \binom{m+1}{k}$, a property of binomial coefficients that is used to build Pascal's triangle.

Are there, for $n \leq 2^{s/2}$, enough pigeons to occupy all pigeonholes? To answer this question we calculate $|\mathcal{N}_s|$, the number of subgraphs of K_n of n vertices that contain a complete graph K_s.

First, we observe that K_n contains $\binom{n}{s}$ distinct complete graphs of s vertices.

Next, we fix a complete graph K_s, a subgraph of K_n, and observe that there are exactly $\binom{n}{2} - \binom{s}{2}$ edges in K_n that do not belong to K_s. This implies that the number of elements of $\mathcal{N}_s$ that contain this fixed graph K_s equals to $2^{\binom{n}{2} - \binom{s}{2}}$.

These two facts, together with Observation 2.3.5, imply

$$\begin{aligned} |\mathcal{N}_s| &= (\text{\# of } K_s \subset K_n) \cdot (\text{\# of subgraphs containing a fixed } K_s) \\ &= \binom{n}{s} \cdot 2^{\binom{n}{2} - \binom{s}{2}} < \frac{n^s}{s!} \cdot 2^{\binom{n}{2} - \binom{s}{2}}. \end{aligned}$$

From $n \leq 2^{\frac{s}{2}}$ we conclude that $n^s \leq 2^{\frac{s^2}{2}}$. This, together with Observation 2.3.6, implies that

$$2n^s \leq 2^{\frac{s^2}{2}+1} = 2^{\frac{s}{2}+1} \cdot 2^{\frac{s(s-1)}{2}} < s! 2^{\binom{s}{2}}$$

or, what is the same, $\frac{n^s}{s!} < \frac{2^{\binom{s}{2}}}{2}$. Therefore,

$$|\mathcal{N}_s| < \frac{n^s}{s!} \cdot 2^{\binom{n}{2} - \binom{s}{2}} < \frac{2^{\binom{s}{2}}}{2} \cdot 2^{\binom{n}{2} - \binom{s}{2}} = 2^{\binom{n}{2}-1} = |\text{PH}|.$$

There are more pigeonholes than pigeons!

Let $H \in \mathcal{N}$ be such that the pigeonhole $\{H, H^c\}$ is empty, i.e. let $H \in \mathcal{N}$ be such that neither H nor its complement H^c contains a complete graph of s vertices. We colour an edge in K_n blue if that edge belongs to H otherwise we colour it red. This colouring avoids a monochromatic K_s. Therefore, $R(s, s) > 2^{\frac{s}{2}}$. □

From Theorem 2.3.2 and Theorem 2.3.3 it follows that, for $s \geq 3$,

$$2^{\frac{s}{2}} < R(s, s) \leq \binom{2(s-1)}{s-1}.$$

The search for tighter bounds for $R(s, s)$ has been ongoing. The lower bound that Erdős established in 1947 was improved by a factor of two by Spencer in 1977.

There has been more progress with the lowering the upper bound. See Conlon's quote at the end of this section.

At this point it seems natural to ask the following question:

Question 2.3.1. Suppose that we decide to use three colours, say blue, red, and green. Is there something like $R(s,t,u)$, for $s,t,u \in \mathbb{N}$? In other words, is it possible to find a number n so that if each edge of K_n is coloured by one of the three colours then there will be always possible to find a blue K_s or a red K_t or a green K_u?

One can go all the way and consider a situation in which, for $m \in \mathbb{N}\backslash\{1\}$, the edges of a complete graph K_n are coloured with m colours.

Definition 2.3.2 (Ramsey Number). *Let $k \in \mathbb{N}\backslash\{1\}$ and $m_1, m_2, \ldots, m_k \in \mathbb{N}\backslash\{1\}$ be given. The Ramsey number $R(m_1, m_2, \ldots, m_k)$ is the minimum number n for which any edge k-colouring of K_n, with colours $c_1, c_2, \ldots, c_k$, contains a c_i-monochromatic K_{m_i}, for some $i \in [1, k]$.*

Question 2.3.1 now becomes:

Question 2.3.2. Does the Ramsey number $R(m_1, m_2, \ldots, m_k)$ exist for any choice of natural numbers $k, m_1, m_2, \ldots, m_k \geq 2$?

We obtain a positive answer to this question as a very special case of Ramsey's theorem. We may apply Theorem 1.3.1 in the setting in which the set of objects that we consider is the set of vertices of a complete graph K_n. We represent, what Ramsey called, a "2-combination" $\{A, B\}$ of two vertices A and B by the corresponding edge, i.e. by the edge in the graph K_n that is incident with the vertices A and B. Ramsey's theorem guarantees that for any partition of the set of edges into a given (finite) number of cells, for n large enough, there will be a set of vertices Δ, of the prescribed size, such that all edges incident with two vertices from Δ belong to the same partition cell.

In other words:

Theorem 2.3.4 (Ramsey's Theorem For Graphs). *For any $k, m_1, m_2, \ldots, m_k \in \mathbb{N}\backslash\{1\}$ the Ramsey number $R(m_1, m_2, \ldots, m_k)$ exists.*

Ramsey numbers and quasi-randomness:

Conlon, who in his doctoral thesis obtained a *super polynomial improvement* on the upper bound of $R(s, s)$, describes his contribution in the following way:

So what I observed is that actually not only does every vertex have the same number of reds and the same number of blues as its neighbours, but actually every pair of vertices has roughly the same number of common red neighbours. And the same is true for their common blue neighbours as well. So it's all very regular. In fact, it's what's called quasi-random. So it looks like a random graph. And knowing that it looks random, actually allows you to compute some of the statistics. So, for example, it allows you to count: instead of triangles, you can now count cliques of size 10. Or you can count cliques of size 100. In fact, you can count cliques of any finite size up to around $\log k$. I was able to count things up to size around $\log k$ over $\log \log k$. Recently, Ashwin Sah [100] was able to count everything up to size $\log k$. But being able to count those cliques allows you to derive a contradiction. And that contradiction, or playing off that contradiction, is what actually gives you the bound in the end. So in some ways, when you ask what my contribution was, my contribution was to see that one could inject quasi-randomness into the situation. And then how to leverage this quasi-randomness [64].

2.4 RAMSEY'S THEOREM: AN INFINITE CASE

"No finite point has meaning without an infinite reference point."

Jean–Paul Sartre, a French philosopher and novelist

1905–1980

In Section 2.3 we restricted our discussion about Ramsey's theorem to finite sets and their subsets of cardinality two. This setting allowed us to formulate a version of Ramsey's theorem in graph-theoretical terms.

There is a red K_s There is a blue K_t

But what if we take an infinite set, the set of natural numbers, for example, and consider a finite partition of the set of all its subsets of cardinality r, for some $r \in \mathbb{N}$? Will we always find an *infinite set* so that all *its* subsets of the cardinality r belong to the same cell of the partition?

Before we proceed with this scenario, we introduce the following notation:

For $r \in \mathbb{N}$ and a set Γ, we define $\Gamma^{(r)}$ to be the set of all subsets on Γ of cardinality r: $\Gamma^{(r)} = \{A \subset \Gamma : |A| = r\}$.

For $k \in \mathbb{N}$ we define a k-colouring of $\mathbb{N}^{(r)}$ as a function from $\mathbb{N}^{(r)}$ to any set of exactly k elements.

If c is a k-colouring of $\mathbb{N}^{(r)}$ and $A \subset \mathbb{N}$ is such that, for all $x, y \in A^{(r)}$, $c(x) = c(y)$, we say that the set A is *monochromatic*.

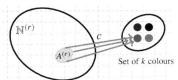

Set of k colours

In this notation, we can re-state Theorem 1.3.1 in the following form:

Theorem 2.4.1 (Ramsey's Theorem). *Let* $k, m, r \in \mathbb{N}$. *There exists* $n \in \mathbb{N}$ *such that whenever* $[1, n]^{(r)}$ *is k-coloured, there is a monochromatic m-set* $M \subseteq [1, n]$.

To fully appreciate the next theorem we observe having arbitrarily large monochromatic sets does not imply the existence of an infinite monochromatic set.

Example 2.4.1 (Infinite vs. Arbitrarily Large). Consider the family of sets of natural numbers:

$$\mathcal{A} = \left\{ A_n = \left[\frac{n(n+1)}{2}, \frac{n(n+3)}{2} \right] : n \in \mathbb{N} \right\} = \{\{1, 2\}, \{3, 4, 5\}, \{6, 7, 8, 9\}, \ldots\}.$$

We define a colouring $c : \mathbb{N}^{(2)} \rightarrow \{\bullet, \blacksquare\}$ by

$$c(\{a, b\}) = \begin{cases} \bullet & \text{if there is } n \in \mathbb{N} \text{ such that } \{a, b\} \subset A_n \\ \blacksquare & \text{otherwise.} \end{cases}$$

For example, what is $c(\{500500, 500501\})$? What about $c(\{499499, 500500\})$?
We are interested in the following three questions:

1. Are there arbitrarily large $\bullet$-monochromatic sets?

2. Is there an infinite $\bullet$-monochromatic set?

3. Is there an infinite $\blacksquare$-monochromatic set?

Solution. To answer the first question, we observe that, for any $n \in \mathbb{N}$ and any $k, \ell \in A_n$, $c(\{k, \ell\}) = \bullet$. In addition

$$|A_n| = \frac{n(n+3)}{2} - \frac{n(n+1)}{2} + 1 = \frac{n}{2} \cdot (n + 3 - n - 1) + 1 = n + 1,$$

which implies that there are arbitrarily large $\bullet$-monochromatic sets.

Let S be any infinite subset of natural numbers. Let $k \in S$ and let $n \in \mathbb{N}$ be such that $k \in A_n$. Since S is an infinite set, there is $\ell \in S$ such that $\ell - k > n + 1$. So ℓ does not belong to A_n and $c(\{k, \ell\}) = \blacksquare$. Therefore, the infinite set S is not $\bullet$-monochromatic.

On the other hand, if we take the set $T = \left\{ \frac{k(k+1)}{2} : k \in \mathbb{N} \right\}$, then no two elements of T belong to the same A_n, $n \in \mathbb{N}$. This means that for any $k, \ell \in T$, $k \neq \ell$, $c(\{k, \ell\}) = \blacksquare$. Hence, T is a $\blacksquare$-monochromatic infinite set. □

Is this always true? Does any two-colouring of $\mathbb{N}^{(2)}$ contain *a* monochromatic infinite set?

The next theorem answers this question.

Theorem 2.4.2 (Ramsey's Theorem for $\mathbb{N}^{(2)}$ – Two Colours). *Whenever $\mathbb{N}^{(2)}$ is two-coloured, there exists an infinite monochromatic set.*

Proof. Let $c : \mathbb{N}^{(2)} \to \{\text{red}, \text{blue}\}$ be a two-colouring.

Our strategy is to build an infinite c-monochromatic set.[2]

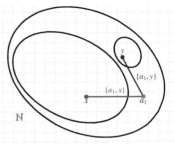

Step 1: Pick $a_1 \in \mathbb{N}$. Look at all red $\{a_1, x\}$ and all blue $\{a_1, y\}$, $x, y \in \mathbb{N}$.

Since the set $\mathbb{N} \setminus \{a_1\}$ is infinite, by the pigeonhole principle at least one of the sets $\{x \in \mathbb{N} : \{a_1, x\}$ is coloured red$\}$ and $\{y \in \mathbb{N} : \{a_1, y\}$ is coloured blue$\}$ is infinite.

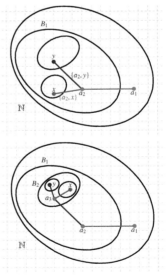

Step 2: Say that $B_1 = \{x \in \mathbb{N} : \{a_1, x\}$ is coloured red$\}$ is infinite. Pick $a_2 \in B_1$. Look at all red $\{a_2, x\}$ and all blue $\{a_2, y\}$, $x, y \in B_1$.

Step 3: Say that $B_2 = \{y \in B_1 : \{a_2, y\}$ is coloured blue$\}$ is infinite. Pick $a_3 \in B_2$. Look at all red $\{a_3, x\}$ and all blue $\{a_3, y\}$, $x, y \in B_2$.

---

[2]We assume the Axiom of Choice.

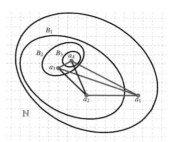

Step 4: Say that $B_3 = \{x \in B_2 : \{a_3, x\}$ is coloured red$\}$ is infinite. Pick $a_4 \in B_3$. Look at all red $\{a_4, x\}$ and all blue $\{a_4, y\}$, $x, y \in B_3$.

We observe that each of the sets $\{a_1, a_2\}, \{a_1, a_3\}, \{a_1, a_4\}$ is coloured red, each of the sets $\{a_2, a_3\}, \{a_2, a_4\}$ is coloured blue, and that the set $\{a_3, a_4\}$ is coloured red.

Continue...

Through this process we obtain an infinite sequence of natural numbers $a_1, a_2, a_3, \ldots$ and an infinite sequence of infinite sets $\mathbb{N} \supseteq B_1 \supseteq B_2 \supseteq B_3 \supseteq \ldots$ with the property that, for any $i \in \mathbb{N}$, $\{a_{i+1}, a_{i+2}, a_{i+3}, \ldots\} \subset B_i$ and

$$c(\{a_i, a_{i+1}\}) = c(\{a_i, a_{i+2}\}) = c(\{a_i, a_{i+3}\}) = \ldots.$$

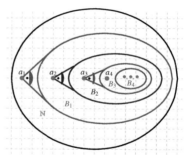

In other words, each a_i sees only red or only blue.

By the pigeonhole principle at least one of the sets $R = \{i \in \mathbb{N} : a_i$ sees only red$\}$ and $B = \{i \in \mathbb{N} : a_i$ sees only blue$\}$ must be infinite.

Say that the set R is infinite. Then the set $\{a_i : i \in R\}$ is an infinite c-monochromatic set. □

Example 2.4.2. Let $a_1, a_2, a_3, \ldots$ be a sequence of mutually distinct real numbers. Prove that it contains a monotone subsequence.

Solution. Let the colouring $c : \mathbb{N}^{(2)} \to \{\bullet, \blacksquare\}$ be defined by

$$c(\{i, j\}) = \begin{cases} \bullet & \text{if } a_{\min\{i,j\}} < a_{\max\{i,j\}} \\ \blacksquare & \text{if } a_{\min\{i,j\}} > a_{\max\{i,j\}}. \end{cases}$$

By Theorem 2.4.2, there is an infinite c-monochromatic set $S = \{i_1 < i_2 < i_3 < \cdots\} \subseteq \mathbb{N}$. Say that, for each $i, j \in S$, $c(\{i, j\}) = \bullet$. By definition of the colouring c, this means that $a_{i_1} < a_{i_2} < a_{i_3} < \cdots$ is a monotone subsequence of the given sequence. □

2.5   RAMSEY'S THEOREM: GENERAL CASE

"The beauty of mathematics only shows itself to more patient followers."

Maryam Mirzakhani, an Iranian mathematician

1977–2017

Theorem 2.4.2 is a special case of the theorem that Ramsey proved in 1930:

Theorem 2.5.1 (Ramsey's Theorem – Infinite Case [92]). *Let* Γ *be an infinite class, and* μ *and* r *positive integers; and let all those sub-classes of* Γ *which have exactly* r *members, or, as we may say, let all* r*-combinations of the members of* Γ *be divided in any manner into* μ *mutually exclusive classes* C_i ($i = 1, 2, 3, \ldots, \mu$)*, so that every* r*-combination is a member of one and only one* C_i; *then, assuming the axiom of selections,* Γ *must contain an infinite sub-class* Δ *such that all the* r*-combinations of the members of* Δ *belong to the same* C_i. [3]

In the notation introduced at the beginning of the previous section, Theorem 2.5.1 can be restated as:

Theorem 2.5.2 (Ramsey's Theorem – Infinite Case). *Let* $k, r \in \mathbb{N}$. *Whenever* $\mathbb{N}^{(r)}$ *is* k*-coloured, there exists an infinite monochromatic set, where, by an infinite monochromatic set* we mean an infinite subset T of $\mathbb{N}$ such that $T^{(r)}$ is monochromatic.

Proof. For a chosen $r \in \mathbb{N}$ let a k-colouring $C : \mathbb{N}^{(r)} \to \{c_1, c_2, \ldots, c_k\}$ be fixed.

The proof is by induction on r. If $r = 1$, then $\mathbb{N}^{(1)} = \{\{n\} : n \in \mathbb{N}\}$. Since C is a finite colouring, by the pigeonhole principle at least one of the colour classes is infinite. Hence, there is $i \in [1, k]$ such that the monochromatic set $\{n \in \mathbb{N} : C(\{n\}) = c_i\}$ is infinite.

Let $r \geq 2$ be such that for any k-colouring of $\mathbb{N}^{(r-1)}$ there exists an infinite monochromatic set. Observe that this assumption implies that if B is any infinite subset of $\mathbb{N}$ and if $B^{(r-1)}$ is k-coloured then B contains an infinite monochromatic subset.[4]

---

[3] This is Ramsey's authentic statement of the theorem. A modern mathematician would likely say *an infinite set* instead of *an infinite class*; *subsets of* Γ instead of *sub-classes of* Γ; *r-subset* instead of *r-combination*; and *the Axiom of Choice* instead of *the axiom of selections*.

[4] Say that $B = \{b_1, b_2, \ldots\}$ and that χ is a k-colouring of $B^{(r-1)}$. Define χ', a k-colouring of $\mathbb{N}^{(r-1)}$, by $\chi'(\{n_1, n_2, \ldots, n_{r-1}\}) = \chi(\{b_{n_1}, b_{n_2}, \ldots, b_{n_{r-1}}\})$.

We build an infinite C-monochromatic set in the following way:[5]

Step 1: Pick $a_1 \in \mathbb{N}$. Let a k-colouring C_1 of $(\mathbb{N} \backslash \{a_1\})^{(r-1)}$ be defined by $C_1(F) = C(F \cup \{a_1\})$. By the induction hypothesis, $\mathbb{N} \backslash \{a_1\}$ contains an infinite C_1-monochromatic set. Call this set B_1.

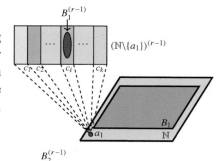

Step 2: Pick $a_2 \in B_1$. Let a k-colouring C_2 of $(B_1 \backslash \{a_2\})^{(r-1)}$ be defined by $C_2(F) = C(F \cup \{a_2\})$. By the induction hypothesis, $B_1 \backslash \{a_2\}$ contains an infinite C_2-monochromatic set. Call this set B_2.

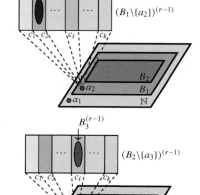

Step 3: Pick $a_3 \in B_2$. Let a k-colouring C_3 of $(B_2 \backslash \{a_3\})^{(r-1)}$ be defined by $C_3(F) = C(F \cup \{a_3\})$. By the induction hypothesis, $B_2 \backslash \{a_3\}$ contains an infinite C_3-monochromatic set. Call this set B_3.

Observe that, for any $F \in B_1^{(r-1)}$, any $F' \in B_2^{(r-2)}$, and any $F'' \in B_3^{(r-3)}$, $C(\{a_1\} \cup F) = C(\{a_1, a_2\} \cup F') = C(\{a_1, a_2, a_3\} \cup F'')$. Also, for any $G \in B_2^{(r-1)}$ and any $G' \in B_3^{(r-2)}$, $C(\{a_2\} \cup G) = C(\{a_2, a_3\} \cup G')$.

Continue. . .

Through this process we obtain an infinite sequence of natural numbers $a_1, a_2, a_3, \ldots$ and an infinite sequence of infinite sets $\mathbb{N} \supseteq B_1 \supseteq B_2 \supseteq B_3 \supseteq \ldots$ with the property that, for any $i \in \mathbb{N}$, $\{a_{i+1}, a_{i+2}, a_{i+3}, \ldots\} \subset B_i$, and, for any $F = \{a_{i_1}, a_{i_2}, \ldots, a_{i_{r-1}}\}, H = \{a_{j_1}, a_{j_2}, \ldots, a_{j_{r-1}}\} \in B_{i+1}^{(r-1)}$, $C(\{a_i\} \cup F\}) = C(\{a_i\} \cup H)$.

---

[5]We assume the Axiom of Choice.

In other words, each a_i sees all $(r-1)$-sets of elements of the sequence that come after it, i.e. all sets $\{a_{j_1}, a_{j_2}, \ldots, a_{j_{r-1}}\} \in B_{i+1}^{(r-1)}$, in the *same* colour.

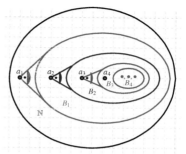

By the pigeonhole principle, at least one of sets $C_j = \{i \in \mathbb{N} : a_i$ sees only the colour $c_j\}$, $j \in [1, k]$, must be infinite. Therefore, there is an infinite C-monochromatic set.

Since C was an arbitrarily k-colouring of $\mathbb{N}^{(r)}$, this completes the induction step and the proof of the theorem. □

We are finally ready to prove Theorem 2.4.1 (and therefore Theorem 1.3.1).

Proof of Theorem 2.4.1. Suppose that the claim of the theorem is not true, i.e. suppose that k, m, and r are such that for any $n \geq r$ there is a k-colouring C_n of $[1, n]^{(r)}$ without a monochromatic m-set. Let $C = \{C_n : n \geq r\}$.

Observe that the number of k-colourings of $[1, r]^{(r)}$ is finite.[6] Since the set C is infinite, by the pigeonhole principle, there is a k-colouring χ_r of $[1, r]^{(r)}$ such that the set $C_r = \{n \geq r : C_n|_{[1, r]^{(r)}} = \chi_r\}$ is infinite.

For $N > r$, inductively, we obtain a k-colouring χ_N of $[1, N]^{(r)}$ such that $\chi_N|_{[1, N-1]^{(r)}} = \chi_{N-1}$ and an infinite set $C_N = \{n \geq N : n \in C_{N-1}$ and $C_n|_{[1, N]^{(r)}} = \chi_N\}$. As a restriction of an infinite number of colourings from the set C, the colouring χ_N is without a monochromatic m-set.

Since $\chi_N|_{[1, N-1]^{(r)}} = \chi_{N-1}$, for all $N > r$, we can define C, a k-colouring of $\mathbb{N}^{(r)}$, in the following way: if $F \in \mathbb{N}^{(r)}$ and if $N = \max F$, then $C(F) = \chi_N(F)$. By Theorem 2.5.2, there is an infinite C-monochromatic set B. But this implies that if M is an m-subset of B then M is χ_N-monochromatic for any $N \geq \max M$, contradicting the fact that χ_N is without a monochromatic m-set. □

Therefore, the claim of the theorem is true. □

The next statement is a refinement of Theorem 2.4.1.

---

[6]Note that the set $[1, r]^{(r)}$ has only one element, the set $[1, r]$. In other words, $[1, r]^{(r)} = \{[1, r]\}$. The number of k-colourings of $[1, r]^{(r)}$ is k.

Corollary 2.5.1 (Ramsey's Theorem – General Finite Form). *For every* $k, r, m_1, m_2, \ldots, m_k \in \mathbb{N}$ *there is* $n \in \mathbb{N}$ *such that, for any* k-*colouring* $C : [1, n]^{(r)} \to \{c_1, c_2, \ldots, c_k\}$, *there are* $i \in [1, k]$ *and a set* $M_i \subseteq [1, n]$ *such that* $|M_i| = m_i$ *and* $C(M_i^{(r)}) = \{c_i\}$.

Proof. Let $m = \max\{m_1, m_2, \ldots, m_k\}$. Let n be a positive integer guaranteed by Theorem 2.4.1 for this particular choice of k, m, and r. Let $C : [1, n]^{(r)} \to \{c_1, c_2, \ldots, c_k\}$ be a k-colouring and let $M \subseteq [1, n]$ be a C-monochromatic m-set. Hence, there is $i \in [1, k]$ such that $C(M^{(r)}) = \{c_i\}$. It follows that for any m_i-set $M_i \subseteq M$, $C(M_i^{(r)}) = \{c_i\}$. □

Definition 2.5.1 (Ramsey Number). *The least number guaranteed by Corollary 2.5.1 is denoted by* $R(r; m_1, m_2, \ldots, m_k)$ *and called a Ramsey number.*

Observe that Ramsey numbers $R(m_1, \ldots, m_k)$, introduced by Definition 2.3.2, correspond to Ramsey numbers $R(2; m_1, \ldots, m_k)$ from the definition above.

Example 2.5.1. Let $a_1, a_2, a_3, \ldots$ be a sequence of real numbers. Prove that it contains either a constant or a strictly monotonic subsequence.

Solution. Let the colouring $c : \mathbb{N}^{(2)} \to \{\bullet, \blacksquare, \blacktriangle\}$ be defined by

$$c(\{i, j\}) = \begin{cases} \bullet & \text{if } a_{\min\{i,j\}} < a_{\max\{i,j\}} \\ \blacksquare & \text{if } a_{\min\{i,j\}} > a_{\max\{i,j\}}. \\ \blacktriangle & \text{if } a_i = a_j. \end{cases}$$

By Theorem 2.5.1, or Theorem 2.5.2, there is an infinite c-monochromatic set $S = \{i_1 < i_2 < i_3 < \cdots\} \subseteq \mathbb{N}$. If, for each $i, j \in S$, $c(\{i, j\}) = \bullet$ or $c(\{i, j\}) = \blacksquare$, then $a_{i_1} < a_{i_2} < a_{i_3} < \cdots$, or $a_{i_1} > a_{i_2} > a_{i_3} > \cdots$, is a monotone subsequence of the given sequence. If, for each $i, j \in S$, $c(\{i, j\}) = \blacktriangle$, then $a_{i_1} = a_{i_2} = a_{i_3} = \cdots$ is a constant subsequence. □

On Ramsey's Theorem:

Graham and Rothschild described Ramsey's theorem in the following way:

The theorem is a profound generalization of the "pigeonhole principle" or "Dirichlet box principle." As is the case with many beautiful ideas in mathematics, Ramsey's theorem extends just the right aspect of an elementary observation and derives consequences which are extremely natural although far from obvious [45].

2.6 EXERCISES

Exercise 2.1 (*Pigeonhole principle and a pair of shoes*). Prove that if there are 10 pairs of shoes on a shelf, picking 11 shoes randomly from the shelf will result in picking up at least one pair of shoes.

Exercise 2.2 (*Pigeonhole principle and a monochromatic rectangle*). Colour each point in the integer grid $[1,82] \times [1,4]$ by •, ■, or ▲. Show that there is a rectangle that has all its vertices of the same colour.

Exercise 2.3 (*Pigeonhole principle and another monochromatic rectangle*). Prove that any three-colouring of the $[1,12] \times [1,12]$ grid yields a monochromatic rectangle.

Exercise 2.4 (*Pigeonhole principle for an ε*). If $a_1, a_2, \ldots, a_{n+1} \in [1,2n]$ are distinct, then there exist i, j, $i \neq j$, such that a_i divides a_j. (This was one of Erdős's favourite questions to ask of an *epsilon*.)

Exercise 2.5 (*Pigeonhole principle and twelve numbers on a circle*). Place the numbers $1, 2, \ldots, 12$ around a circle, in any order. Prove that there are three consecutive numbers which sum to at least 19.

Exercise 2.6 (*Pigeonhole principle and monotone sequences*). You program your computer to randomly pick, one by one, 50 positive integers. This generates a sequence $a_1, a_2, \ldots, a_{50}$, where the index i means that the integer a_i was the ith randomly picked positive integer. Observe that, since the whole process is random, the sequence $a_1, a_2, \ldots, a_{50}$ may not be ordered. In other words, for any $i \leq 49$, a_{i+1} may be greater than or equal to or less than a_i.

After generating several sequences, you notice that each time you can find at least 8 members of the sequence, say $a_{i_1}, a_{i_2}, \ldots, a_{i_8}$, that form a non-decreasing subsequence, i.e. for each $j \in [1,7]$, $a_{i_{j+1}} \geq a_{i_j}$, OR that they form a non-increasing subsequence, i.e. for each $j \in [1,7]$, $a_{i_{j+1}} \leq a_{i_j}$.

You wonder if this is just a coincidence or it is true that something like this must always happen. What would you do?

Note: Actually, it is true that any set of $n^2 + 1$ positive integers, there is a non-decreasing or a non-increasing subsequence of length $n+1$. This fact was established by Erdős and Szekeres in 1935 [28]. Can you prove this statement?

Exercise 2.7 (*Pigeonhole principle and the number of friends*). You are attending a *Happy New School Year* party. At some point a person next to you says: "Do you know that at this party there must be at least two people with the exactly same number of friends among the party attendees?"

You are puzzled: "How would you prove something like that?"

Exercise 2.8 (*Pigeonhole principle and consecutive integers*). Let A be a set of an odd number of consecutive positive integers. Say, $A = \{k, k+1, \ldots, k+2n\}$ for some $k, n \in \mathbb{N}$. Let B be any subset of the set A that contains at least $n+1$ elements. Prove that there are $a, b \in B$ such that $a + b = 2k + 2n$.

Exercise 2.9 (*Pigeonhole principle and digits and divisibility*). Consider the set $A = \{1, 11, 111, 1111, \ldots\}$, the set that contains all natural numbers whose decimal representation uses only the digit 1. Prove that A contains an element that is divisible by 2023.

Exercise 2.10 (*Pigeonhole principle and the density of a set*). For a positive number x we define the *fractional part* of x as $\{x\} = x - \lfloor x \rfloor$. Observe that $\{x\} \in [0, 1)$.

Let α be a positive irrational number. Use the pigeonhole principle to prove that the set $F_\alpha = \{\{n\alpha\} : n \in \mathbb{N}\}$ is dense in the interval $[0, 1]$, i.e. prove that, regardless how small it is, any interval $(a, b) \subset (0, 1)$ contains an element from F_α. *Note:* Recall that, since α is an irrational number, $0 \notin F_\alpha$.

Exercise 2.11 (*Pigeonhole principle and Ramsey's theorem*). Recall Theorem 1.3.1, the formulation of Ramsey's theorem as it was stated by Frank Ramsey in 1930.

1. By choosing appropriate values of r, n, and μ, show that the pigeonhole principle, Theorem 2.1.1, is a special case of Ramsey's theorem.

2. By choosing appropriate values of r, n, and μ, show that the generalized pigeonhole principle, Theorem 2.1.2, is a special case of Ramsey's theorem.

Exercise 2.12 (*Ramsey's theorem and independent graphs*). An independent subgraph of a given graph consists of vertices of which no pair is adjacent. Prove that any graph with more than six vertices contains a complete subgraph on three vertices or an independent subgraph of three vertices.

Exercise 2.13 (*How many monochromatic triangles are there?*). True or False: each two-edge colouring of K_6 yields at least *two* monochromatic triangles?

Exercise 2.14 (*Ramsey number $R(4,4)$ is greater than* 17). In this problem you will show that there is a two-colouring of K_{17} with no monochromatic K_4.

Label each vertex of K_{17} by one of the integers $0, 1, 2, \ldots, 16$.

Let $\langle x, y \rangle$ denote the edge between the vertices $x, y \in [0, 16]$, with $x < y$.

We define a two-colouring of the edges of K_{17} in the following way:

$$C(\langle x, y\rangle) = \begin{cases} \bullet & \text{if } y - x \in \{1,2,4,8,9,13,15,16\} \\ \blacksquare & \text{if } y - x \in \{3,5,6,7,10,11,12,14\}. \end{cases}$$

1. Prove that any K_4 that contains the edge $\langle 0, 1\rangle$ is not C-monochromatic.

2. Prove that for any $x, y \in [0, 16]$, $x < y$, $C(\langle x, y\rangle) = C(\langle 0, y - x\rangle)$. Explain why this fact implies that to check if K_4 with vertices x, y, z, w, $x < y < z < w$, is monochromatic, it is enough to check if K_4 with vertices $0, y - x, z - x, w - x$ is monochromatic.

3. Recall that you have established that any K_4 that contains the edge $\langle 0, 1\rangle$ is not C-monochromatic. So, let us consider the complete graph K_4 with the vertices $0, x, y, z$, with $2 \le x < y < z$.

   (a) Suppose that $C(\langle 0, x\rangle) = C(\langle 0, y\rangle) = C(\langle 0, z\rangle) = \bullet$. Show that, under this condition, K_4 is not C-monochromatic.

      i. Show that $x \in \{2, 4, 8, 9, 13\}$, $y \in \{4, 8, 9, 13, 15\}$ and $z \in \{8, 9, 13, 15, 16\}$.
      ii. Show that if $x \in \{2, 9, 13\}$ then $C(\langle x, y\rangle) = \blacksquare$ or $C(\langle x, z\rangle) = \blacksquare$ or $C(\langle y, z\rangle) = \blacksquare$. Conclude that K_4 is not monochromatic.
      iii. Show that if $x \in \{4, 8\}$ and if $C(\langle x, y\rangle) = C(\langle x, z\rangle) = \bullet$ then $C(\langle y, z\rangle) = \blacksquare$. Conclude that K_4 is not monochromatic.

   (b) Suppose that $C(\langle 0, x\rangle) = C(\langle 0, y\rangle) = C(\langle 0, z\rangle) = \blacksquare$. Show that, under this condition, K_4 is not C-monochromatic.

      i. Show that $x \in \{3, 5, 6, 7, 10, 11\}$, $y \in \{5, 6, 7, 10, 11, 12\}$, and $z \in \{6, 7, 10, 11, 12, 14\}$.
      ii. Show that if $x \in \{10, 11\}$ then $C(\langle x, y\rangle) = \bullet$ and conclude that K_4 is not monochromatic.
      iii. Show that if $x \in \{3, 5, 6, 7\}$ and if $C(\langle x, y\rangle) = C(\langle x, z\rangle) = \blacksquare$ then $C(\langle y, z\rangle) = \bullet$ and conclude that K_4 is not monochromatic.

Exercise 2.15 (*Ramsey's theorem and a question about a contradiction*). In Exercise 2.14 we established that there is a two-colouring of K_{17} with no monochromatic K_4. Does this contradict the fact that $R(4,3) = R(3,4) = 9$? Why yes, or why not?

Exercise 2.16 (*Ramsey number $R(3,3,3)$ is less than or equal to 17*). Show that every three-colouring of the edges of K_{17} contains a monochromatic K_3.

Exercise 2.17 (*Ramsey number $R_r(3)$ is less than or equal to $3r!$*). For $r \in \mathbb{N}\setminus\{1\}$, prove that $R_r(3) = R(\underbrace{3, 3, \ldots, 3}_{r}) \leq 3r!$

Exercise 2.18 (*Ramsey's theorem and an infinite monochromatic set*). Recall that Ramsey's theorem guarantees that whenever $\mathbb{N}^{(r)}$ is two-coloured, there exists an infinite monochromatic set.

Let the colouring $C : \mathbb{N}^{(3)} \to \{\bullet, \blacksquare\}$ be defined by

$$C(\{x < y < z\}) = \begin{cases} \bullet & \text{if } x \mid (z - y) \\ \blacksquare & \text{otherwise.} \end{cases}$$

Find an infinite C-monochromatic set.

Exercise 2.19 (*Ramsey's theorem, infinite case, and collinear points*). Let S be an infinite set of points in the plane. Show that there is an infinite subset A of S such that either no three points of A are on a line, or all points of A are on a line.

Exercise 2.20 (*Ramsey's theorem, infinite case, and a question if a function is convex or concave*). Let y_1, y_2, y_3, ... be a sequence of mutually distinct real numbers. Use Ramsey's theorem to prove that the sequence $(1, y_1)$, $(2, y_2)$, $(3, y_3)$, ... of points in $\mathbb{R}^2$ contains a subsequence such that the induced function is convex or concave.

Exercise 2.21 (*Ramsey's theorem, infinite case, and a question if a function is constant, convex, or concave*). Let $y_1, y_2, y_3, \ldots$ be a sequence of real numbers. Use Ramsey's theorem to prove that that the sequence $(1, y_1)$, $(2, y_2)$, $(3, y_3)$, ... of points in $\mathbb{R}^2$ contains a subsequence such that the induced function is constant, convex, or concave.

van der Waerden's Theorem

van der Waerden's THEOREM is another of the bedrocks of Ramsey theory. The theorem establishes as a fact that any finite colouring of natural numbers contains monochromatic arithmetic progressions of any (finite) length.

We start this chapter with a section about Bartel Leendert van der Waerden's life and work. This is followed by a visualized proof of van der Waerden's theorem. We reflect about several questions and results that sprout from the theorem and have marked the development of Ramsey theory.

The end of the chapter includes exercises that may help the reader to better grasp some of the topics presented in this chapter.

3.1 BARTEL VAN DER WAERDEN

"[My father] maintained that I should play outside rather than dedicate myself to mathematics books."

Bartel Leendert van der Waerden

3.1.1 Who Was Bartel Leendert van der Waerden?

Bartel Leendert van der Waerden was a Dutch mathematician and historian of mathematics and science.

DOI: 10.1201/9781003286370-3

Bartel Leendert van der Waerden made contributions to:

abstract algebra	algebraic geometry	analysis
combinatorics	geometry	group theory
history of ancient science	history of astronomy	history of mathematics
history of modern physics	mathematical statistics	number theory
probability theory	quantum mechanics	topology

Birth and Death. Bartel Leendert van der Waerden was born in Amsterdam, Netherlands, on February 2, 1903, and died on January 12, 1996, in Zürich, Switzerland.

Timeline: Ramsey – van der Waerden – Erdős

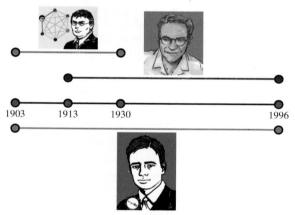

Bartel's Family. Bartel van der Waerden's parents were Dorothea Adriana Endt and Theodorus van der Waerden. Theodorus van der Waerden was a teacher of mathematics and mechanics. Bartel had two younger brothers, Coenraad and Benno. In 1929 he married Camilla Rellich. Camilla and Bartel had three children, Helga, Ilse, and Hans.

Bertel discovered his mathematical talents and interests early. Many years later, van der Waerden recalled that at the young age he "rediscovered all of trigonometry, starting from the law of cosines [23]."

So it should not be a surprise that van der Waerden, still in his early twenties, was accepted in the circles of some of the leading mathematicians of the time. For example, in 1925, Emmy Noether, a German mathematician, 1882–1935, wrote to one of van der Waerden's former professors: "That van der Waerden would give us much pleasure was correctly foreseen by you. The paper he submitted in August to the *Annalen* is most excellent [121]."

Between 1931–1945 van der Waerden was a professor of mathematics at the University of Leipzig, Germany. During the rise of the Third Reich and through World War II, van der Waerden remained at Leipzig and continued with his work. This part of van der Waerden's life was described and discussed in, for example, [109, 110, 121].

- At the peak of their activity, between the outbreak of World War I in 1914 and the Nazis' rise to power in 1933, one-third of all math professors in Germany were Jewish – although Jews constituted less than 1 percent of the total population. These mathematicians served on the editorial boards of leading academic journals and were involved in the founding of the mathematical society.

- Of the 90 Jewish mathematicians chronicled in a recent historic study, three committed suicide after the Nazis rose to power and two were killed in the Holocaust. The rest managed to emigrate.

- The situation was particularly dire at Göttingen: Three out of four of the heads of the university's mathematics and physics institutes had been Jews. Not long after the mass expulsion, a reception was held at the university, at which Nazi education minister Bernhard Rust met the former director of the mathematics institute. Rust asked him if it had been harmed by the expulsion of the Jews. "It has not been harmed, sir," replied the former director. "It has simply ceased to exist."

Ofer Aderet: "Setting the record straight about Jewish mathematicians in Nazi Germany"
Haaretz, November 25, 2011

After the war, for several years van der Waerden lived and worked in the Netherlands and the United States. In 1951, van der Waerden was appointed as a professor of mathematics at the University of Zürich. He lived in Zürich for the rest of his life.

In his wife Camilla's words, Bartel van der Waerden "was always a great solitary figure."

3.1.2   van der Waerden's Work: Two Examples

3.1.2.1   *History of Mathematics*

In an interview that he gave in 1993, van der Waerden recalled that he was interested in history of mathematics since his time as an undergraduate student

[23]. In addition, throughout his career, van der Waerden was interested in history of astronomy and history of science in general. For example, van der Waerden studied and extensively wrote about various aspects of Egyptian, Babylonian, Greek, and Indian astronomy.

van der Waerden was a mathematician who, with his research results, was making mathematical history himself. Hence, a modern reader could be curious why he would be so interested in how and why ancient mathematicians, astronomers, and philosophers understood and described the mathematical and scientific side of the world as they saw it. A comment by Camilla van der Waerden may help: "[History of mathematics and astronomy] pleased him the most, to tell the truth, for many years."

van der Wearden's historical writings include the book *A History of Algebra from al–Khwārizmī to Emmy Noether* published in 1985 [124]. It should be mentioned that a large body of van der Waerden's own research was in the field of algebra.

The book starts with Muhammad ibn Mūsā al–Khwārizmī, a Persain mathematician, c. 280–c. 850, from whose phrase "on the solution of problems by *al-jabr* and *al-muqabala*" the word "algebra" was derived.

For van der Waerden, the work of Évariste Galois, a French mathematician, 1811–1832, was the watershed moment in the development of algebra. In van der Waerden's words:

> Modern algebra begins with Évariste Galois. With Galois, the character of algebra changed radically. Before Galois, the efforts of algebraists were mainly directed towards the solution of algebraic equations. After Galois, the efforts of the leading algebraists were mainly directed towards the investigation of the structure of rings, fields, algebras, and the like.

Muhammad ibn Mūsā al–Khwārizmī, Évariste Galois, and Emmy Noether, three trailblazers in the history of algebra

The account of the work of Emmy Noether spans over 40 pages of van der Waerden's book. This is a segment of the book where the author is not just a historian, but also one who provides his own testimonial about the heroine of this period of the history of algebra:

> In 1924, when I came to Göttingen as a student, I had the pleasure to attend a course of Emmy Noether on Hypercomplex Numbers. In 1926/27 she again lectured on the same subject. This time the title of her course was "Hyperkomplexe Zahlen und Darstellungstheorie." My lecture notes are lost, but the contents of my notes were incorporated in Emmy Noether's 1929 paper *Hyperkomplexe Größen und Darstellungstheorie*, Math. Zeitschrift 30, pages 641–692.

3.1.2.2 van der Waerden's Theorem

Considered as "one of the most elegant pieces of mathematics ever produced" [20], together with Hilbert's cube lemma, Schur's theorem, and Ramsey's theorem, van der Waerden's theorem is one of the pillars of Ramsey theory.

Theorem 3.1.1 (van der Waerden's Theorem [122].). *For any given positive integers k and l, there is some number $n = n(l,k)$ such that if the set of integers $[1,n]$ is l-coloured then there is a k-term monochromatic arithmetic progression.*

Vocabulary of van der Waerden's theorem:

- A k-term arithmetic progression is any set of the form $\{a + jd : j \in [0, k-1]\}$, where $a, d \in \mathbb{R}$, $d \neq 0$. We will write this arithmetic progression as $a, a + d, a + 2d, \ldots, a + (k-1)d$. We say that d is the common difference of the arithmetic progression.

- For a given $l \in \mathbb{N}$, an l-colouring of a set A is any function $c : A \to C$, where the set C has exactly l elements. For example, one may take $C = [1, l]$. We will say that each element of the set C is a colour.

- Let $c : A \to C$ be an l-colouring of the set A and let $B \subseteq A$. We say that the *set B is c-monochromatic* if, for any $x, y \in B$, $c(x) = c(y)$. We omit the explicit reference to the colouring c, if it is clear from the context.

- Let $c : \mathbb{N} \to C$ be a colouring of the set of positive integers. We say that the arithmetic progression $a, a + d, a + 2d, \ldots, a + (k-1)d$ is monochromatic if $c(a) = c(a+d) = \cdots = c(a + (k-1)d)$.

---

van der Waerden's theorem was proved in 1926 and published in 1927. Many years later, van der Waerden [123] told a story about how the proof was found.[1] See below for an excerpt from a short graphic story [57] inspired by van der Waerden's essay.

The beginning of a graphic short story "How the proof of Baudet's conjecture was found."

In his essay, van der Waerden provided insight into how the proof was created and reflected on the process of mathematical discovery.

(...) One of the main difficulties in the psychology of invention is that most mathematicians publish their results with condensed proofs, but do not tell us how they found them. In many cases they do not

---

[1]For the whole essay *How the proof of Baudet's conjecture was found*, see also [109], pages 310–318.

even remember their original ideas. Moreover, it is difficult to explain our vague ideas and tentative attempts in such a way that others can understand them.

(...) In the case of our discussion of Baudet's conjecture the situation was much more favourable for a psychological analysis. All ideas we formed in our minds were at once put into words and explained by little drawings on the blackboard. We represented the integers $1, 2, 3, \ldots$ in two classes by means of vertical strokes on two parallel lines. Whatever one makes explicit and draws is much easier to remember and to reproduce than mere thoughts.

(...) This final idea was accompanied by a feeling of complete certainty. I cannot explain this feeling; I can only say that the mathematicians often have such a conviction. When a decisive idea comes to our mind, we feel that we have the whole proof we are looking for: we have only to work it out in detail.

Regardless of the fact that combinatorics was "a field that he never seriously worked in" [125], van der Waerden's contribution to combinatorics is indispensable. Various generalizations of van der Waerden's theorem have marked the development of Ramsey theory over the last several decades. In Section 3.5 we will mention a few of those generalizations.

3.2 VAN DER WAERDEN'S THEOREM: THREE-TERM ARITHMETIC PROGRESSIONS

"Say what you know, do what you must, come what may."

Sofia Vasilyevna Kovalevskaya, a Russian
mathematician

1850–1891

As an introduction to our investigation of van der Waerden's theorem, we start this section with two examples.

Example 3.2.1. Complete each of two-colourings in the figure to the right, by using colours ⊙ and ▣, so that the new colouring avoids monochromatic three-term arithmetic progressions.

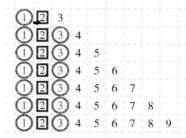

And, yes, it is impossible to colour the set [1,9] with two colours and avoid a monochromatic three-term arithmetic progression! We will refer to this fact throughout this chapter.

Example 3.2.2. Check if the three-colouring of the set [1, 17] below avoids monochromatic four-term arithmetic progressions:

①②③④⑤⑥⑦⑧⑨⑩⑪⑫⑬⑭⑮⑯⑰

Is it possible to colour, by using the already given colours, numbers 18 and 19 to avoid a monochromatic four-term arithmetic progression?

①②③④⑤⑥⑦⑧⑨⑩⑪⑫⑬⑭⑮⑯⑰ 18 19

It seems natural to ask the following questions:

Question 3.2.1. Is it possible to extend the three-colouring from Example 3.2.2 to the interval [1, 25], so that the new colouring avoids monochromatic four-term arithmetic progression? What about interval [1, 50]? [1, 100]? ℕ?

Also, do answers to this kind of questions depend on the number of colours? The length of the monochromatic arithmetic progression?

It is possible that Pierre Joseph Henry Baudet, a Dutch mathematician, 1891–1921, considered a similar set of questions when he stated his celebrated conjecture.

Conjecture 3.2.1 (Baudet's Conjecture [122]). *If the sequence of integers 1, 2, 3, . . . is divided into two classes, at least one of the classes contains an arithmetic progression of k terms, no matter how large the length k is.*

As we have already learned, in 1926 in Hamburg, Germany, over a lunch, van der Waerden introduced Baudet's conjecture to Emil Artin, an Austrian mathematician, 1898–1962, and Otto Schreier, a Jewish–Austrian mathematician, 1901–1929. The three young mathematicians would spend the rest of the day in the front of a blackboard discussing the conjecture and coming up with some of the main ingredients of the proof. In the following year, van der Waerden would publish a paper entitled *Beweis einer Baudetschen Vermutung* (*Proof of Baudet's Conjecture*) and establish the famous fact.

Actually, van der Waerden proved a bit more than Baudet had conjectured. As we stated in Theorem 3.1.1, for given $k, l \in \mathbb{N}$, there exists a positive integer $n = n(l, k)$ with the property that if the set of integers $[1, n]$ is l-coloured then there is a k-term monochromatic arithmetic progression.

This means that we do not need to colour *all* natural numbers with a provided set of l colours to be sure that, somewhere, there is a monochromatic arithmetic progression of the prescribed length k.

Definition 3.2.1 (van der Waerden Number). *The smallest $n = n(l, k)$ guaranteed by Theorem 3.1.1 is called the van der Waerden number $W(l; k)$.*

Recall that in Example 3.2.1 we established that $W(2; 3) = 9$. Also, since it is known that $W(3; 4) = 293$, the answer to Question 3.2.1 would be that any extension of the given three-colouring to the interval $[1, 293]$ will contain a monochromatic four-term arithmetic progression.

In the remainder of this section we introduce the main tools, terminology, and ideas that we will use to prove Theorem 3.1.1. To illustrate how those tools and ideas work, we will first apply them in the case $k = 3$, i.e. we will start by looking for three-term monochromatic arithmetic progressions in any given finite colouring of a certain subset of the set of natural numbers.

3.2.1 Colour-focused Arithmetic Progressions and Spokes

We start by defining what it means that a set of arithmetic progressions is colour-focused, a term used by Leader [74].

Definition 3.2.2 (Colour-focused Arithmetic Progressions). *Let c be a finite colouring of an interval of positive integers $[\alpha, \beta]$ and let k and r be positive integers. We say that k-term arithmetic progressions $A_1, A_2, \ldots, A_r$, where*

$$A_i = \{a_i + j d_i : j \in [0, k-1]\} \subseteq [\alpha, \beta], i \in [1, r],$$

are colour-focused at a positive integer f if:

1. *Each A_i is monochromatic.*

2. *If $i \neq j$ then A_i and A_j are not of the same colour.*

3. *$a_1 + k d_1 = a_2 + k d_2 = \cdots = a_r + k d_r = f$.*

We will say that f is the focus of arithmetic progressions $A_1, A_2, \ldots, A_r$.
Each $(k+1)$-term arithmetic progression $A_i \cup \{f\}$, $i \in [1, r]$, is called a spoke.

Example 3.2.3. Suppose that c is a two-colouring of the set $[1, 6]$ and that $c(1) = c(4) = \bullet$ and $c(3) = c(5) = \blacksquare$. This means that two-term arithmetic progressions $A_1 = \{1, 1 + 3 = 4\}$ and $A_2 = \{3, 3 + 2 = 5\}$ are monochromatic.

From $1 + 2 \cdot 3 = 7$ and $3 + 2 \cdot 2 = 7$, by definition, the arithmetic progressions A_1 and A_2 are colour-focused at 7.

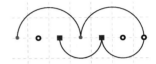

In the figure to the right we visualize a *polychromatic fan*, the term that Tao uses for a set of colour-focused arithmetic progressions [119]. This visualization also justifies the term *spoke* used in Definition 3.2.2.

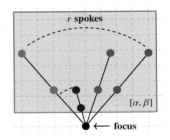

3.2.2 van der Waerden's Theorem for Three-term Arithmetic Progressions

We fix the length of monochromatic arithmetic progressions to $k = 3$ to obtain a special case of Theorem 3.1.1:

Theorem 3.2.1 (van der Waerden's Theorem, $k = 3$). *Any l-colouring of the positive integers contains a monochromatic three-term arithmetic progression. Moreover, there is a natural number n such that any l-colouring of the interval of positive integers $[1, n]$ contains a monochromatic three-term arithmetic progression.*

As a warm-up, we start with the case of two colours.

Proof of Theorem 3.2.1 in the case $l = 2$. Recall that there are $2^3 = 8$ distinct two-colourings of the set $[1, 3]$.

Since we treat both colours equally, and since by the pigeonhole principle there must be a monochromatic set with at least two elements, the figure below shows all different possibilities of colouring of the set $[1,3]$ with two colours.

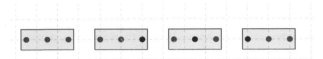

We observe that any two-colouring of the set $[1,3]$ produces either a monochromatic three-term arithmetic progression *or* one colour-focused two-term arithmetic progression. Also, we observe that in the case of a colour-focused two-term arithmetic progression, the focus must belong to the set $[1,5]$.

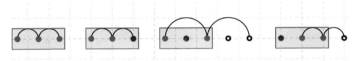

Next we consider the interval of positive integers $[1,5 \cdot (2^5 + 1)] = [1,165]$. We divide this interval into 33 consecutive blocks B_i, $i \in [1,33]$, of length 5. See below.

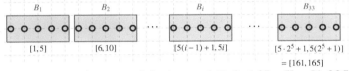

Let c be a two-colouring of the interval $[1,2 \cdot 165 - 5] = [1,325]$.

Recall that there are $2^5 = 32$ distinct two-colourings of five (fixed) consecutive integers. By the pigeonhole principle, there are two blocks, B_i and B_j, $1 \le i < j \le 33$, that are identically coloured by the colouring c.

If the block B_i contains a c-monochromatic three-term arithmetic progression, we are done.

Otherwise B_i (and B_j) contain one colour-focused two-term arithmetic progression and we have a situation like one given in the figure.

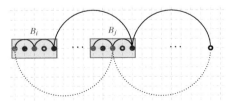

Observe that, since $B_i \cup B_j \subset [1,165]$ and since the difference between the corresponding elements in these two blocks can be at most 160, the common focus of two colour-focused two-term arithmetic progressions represented in the figure must belong to the interval $[1, 325]$. Since c is a two-colouring, one of the spokes must be a c-monochromatic three-term arithmetic progression.

Therefore, every two-colouring of $[1, 325]$ contains a monochromatic three-term arithmetic progression. In other words, $W(2; 3) \leq 325$. □

As van der Waerden described in [123], and as it is shown in the frames from the graphic story [57], the above line of thinking was part of the conversation that he, Artin, and Schreier had on that fateful day in 1926 in Hamburg, when they discussed Baudet's conjecture.

Our next step is to consider an l-colouring of a certain finite set of natural numbers, for any $l \geq 2$.

Proof of Theorem 3.2.1 in the case $l \geq 2$. Let $l \geq 2$ be an integer. The strategy of the proof is to use mathematical induction on $r \in [1, l]$ to prove the claim below and then to observe that, in the case $r = l$, the claim all but completes the proof of Theorem 3.2.1.

Claim 3.2.1. *For all $r \in [1, l]$, there exists a natural number M such that whenever $[1, M]$ is l-coloured, either there exists a monochromatic three-term arithmetic progression or there exist r colour-focused arithmetic progressions of length 2.*

Proof of Claim 3.2.1. To establish the base step we take $r = 1$ and $M = l + 1$.

If the interval $[1, l + 1]$ is l-coloured then there is a monochromatic three-term arithmetic progression or, by the pigeonhole principle, at least *one* colour-focused two-term arithmetic progression.

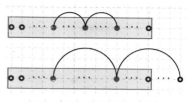

For the induction step, let $r \in [2, l]$ have the property that there is an $m \in \mathbb{N}$ such that any l-colouring of $[1, m]$ contains a monochromatic three-term arithmetic progression or $r - 1$ spokes, i.e. $r - 1$ colour focused two-term arithmetic progressions.

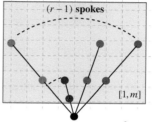

Next, we consider the interval $[1, 2m(l^{2m} + 1)]$. We divide $[1, 2m(l^{2m} + 1)]$ into $l^{2m} + 1$ consecutive blocks of length $2m$.

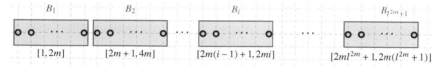

Let c be an l-colouring of the interval $[1, 2m(l^{2m} + 1)]$. Suppose that c does not contain a monochromatic three-term arithmetic progression.

This implies, by the induction hypothesis, i.e. by the choice of the positive integer m, that each block B_i contains $r - 1$ spokes together with their focus.

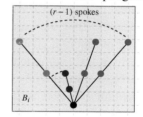

Since there are l^{2m} distinct l-colourings of any fixed $2m$ consecutive integers and since there are $l^{2m} + 1$ blocks, by the pigeonhole principle, there must be two blocks, B_i and B_j, $i < j$, coloured in the same way. We can think about the coloured block B_j as a translate of the coloured block B_i.

Both B_i and B_j contain $r-1$ spokes together with their focus. Observe how, in the interval $[1, 2m(l^{2m}+1)]$, r new spokes with the common focus emerge.

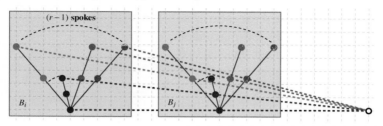

A closer look: a pair of corresponding spokes in B_i and B_j, i.e. a spoke in B_i and its translate in B_j, produce a new pair of spokes in $[1, 2m(l^{2m}+1)]$.

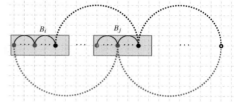

The emergence of r spokes in $[1, 2m(l^{2m}+1)]$, establishes that the number $M = 2m(l^{2m}+1)$ has the property that any l-colouring of $[1, M]$ contains a monochromatic three-term arithmetic progression or r spokes. This completes the induction step and the proof of Claim 3.2.1. □

We complete the induction step in the proof of Theorem 3.2.1 by considering an l-colouring of the interval $[1, 2M]$, where M is the positive integer guaranteed by Claim 3.2.1 in the case $r = l$.

If we assume that the restriction of this colouring on interval $[1, M]$ does not contain a three-term monochromatic arithmetic progression then, by Claim 3.2.1, it must contain l spokes with the common focus in $[1, 2M]$. Since we have l spokes and l colours, by the pigeonhole principle, one of those spokes must be a three-term monochromatic arithmetic progression.

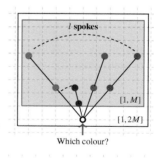

Which colour?

Hence, $W(l; 3) \leq 2M$.

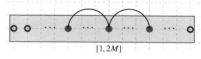

□

About $W(l;3)$:

The reader has probably noticed that in Example 3.2.2 we established that $W(2;3) = 9$ and that the presented proof of Theorem 3.2.1 implied that $W(2;3) \leq 325$. One may wonder about the merit of finding an upper bound that is more than 35 times larger than the true value. The main reason that makes this upper bound so important is that it is obtained by a method that provided us with insight of how to actually prove Theorem 3.2.1. As we will see in the next section, the same approach may be used to prove van der Waerden's theorem for any finite number of colours l and any k-term arithmetic progression.

This is a common scenario in Ramsey theory: we know that a number with the certain property exists, but since we do not know its exact value, improving the existing bounds becomes a research question on its own. In such a scenario, even in "small" cases the search for the exact value, or for a narrow interval that contains the exact value, may become a challenging computational problem. For the general case, our only hope is that the amount of the accumulated knowledge and a moment of human brilliance will produce a breakthrough.

At the time of writing this text, the exact values of $W(l;3)$ are known only for $l \in \{2,3,4\}$: $W(2;3) = 9$, $W(3;3) = 27$, and $W(4;3) = 76$.

3.3 PROOF OF VAN DER WAERDEN'S THEOREM

"Mathematics is really there for you to discover."

Ron Graham

This section contains a proof of van der Waerden's theorem. The proof follows van der Waerden's original approach to establish the existence of the number $n(l,k)$ by using double induction. As Nicolaas Govert de Bruijn, a Dutch mathematician, 1918–2012, said: "van der Waerden's argument is so nice that one might secretly hope that a simpler proof does not exist!" [20].[2]

We will continue to use the tools and terminology introduced in Section 3.2.1.

Proof of Theorem 3.1.1. The strategy of the proof is to use mathematical induction on k, the length of a monochromatic arithmetic progression, to prove that $n(l,k)$ exists for any number of colours $l \in \mathbb{N}$.

---

[2]There are several other proofs of van der Waerden's theorem. See, for example, [3, 22, 40, 44, 81].

For the base case observe that, for any positive integer l, $W(l; 1) = 1$ and $W(l; 2) = l + 1$. Also, recall that, by Theorem 3.2.1, we know that $W(l; 3)$ exists for all $l \in \mathbb{N}$.

For the induction step we consider $k \geq 4$ such that $W(l; k - 1)$ exists for any $l \in \mathbb{N}$. We fix $l \geq 2$.

To verify the induction step we will need the following claim:

Claim 3.3.1. *For all $r \in [1, l]$, there exists a natural number M such that whenever $[1, M]$ is l-coloured, either there exists a monochromatic k-term arithmetic progression or there exist r colour-focused $(k - 1)$-term arithmetic progressions.*

Proof of Claim 3.3.1. We use mathematical induction on $r \in [1, l]$.

For the base case $r = 1$ we take $M = 2W(l; k - 1)$.

Any l-colouring of the set $[1, M]$ produces either a monochromatic k-term arithmetic progression

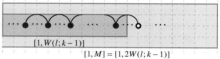

or at least *one* colour-focused $(k - 1)$-term arithmetic progression.

For the induction step we take $r \in [2, l]$ such that there is an m such that any l-colouring of $[1, m]$ contains a monochromatic k-term arithmetic progression or $r - 1$ spokes, i.e. $r - 1$ colour-focused $(k - 1)$-term arithmetic progressions.

Where are you?

> **The base case:**
> For any l, $W(l; 1) = 1$, $W(l; 2) = l + 1$, $W(l; 3)$ **exists**
>
> **The induction step:** The induction hypothesis is
> that k is such that $W(l; k - 1)$ exists for any l
>
> > Fix l. **Claim: For all $r \leq l$, there exists a natural number M**
> > **such that whenever $[1, M]$ is l-coloured, [...]**
> >
> > > **The base case:** $r = 1$
> > >
> > > **The induction step:**
> > >
> > > **You are HERE!**

Observe that, by our assumption, any l-colouring of any interval of $2m$ consecutive integers $[a, a + 2m - 1]$, contains a monochromatic k-term arithmetic progression or at least $r - 1$ colour-focused $(k - 1)$-term arithmetic progressions focused at some $f \in [1, 2m]$.

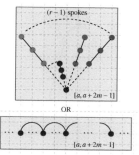

Recall that, by the induction hypothesis, the van der Waerden number $W(l^{2m}; k - 1)$ exists.[3]

We consider the interval $[1, 2mW(l^{2m}; k - 1)]$ and divide this interval into $W(l^{2m}; k - 1)$ consecutive blocks B_i, $1 \leq i \leq W(l^{2m}; k - 1)$, of length $2m$.

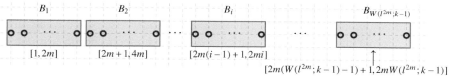

Suppose that c is an l-colouring of $[1, 2mW(l^{2m}; k - 1)]$ that does not contain a monochromatic k-term arithmetic progression. The restriction of c on a block B_i, $i \in [1, W(l^{2m}; k - 1)]$, is one of the l^{2m} possible l-colourings of B_i.

It follows that the l-colouring c of $[1, 2m \cdot W(l^{2m}; k - 1)]$ induces an l^{2m}-colouring of $[1, W(l^{2m}; k - 1)]$.

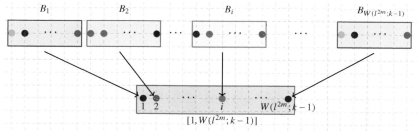

By the definition of the van der Waerden number, any l^{2m}-colouring of $[1, W(l^{2m}; k - 1)]$ contains a monochromatic $(k - 1)$-term arithmetic progression. Consequently, the l^{2m}-colouring of $[1, W(l^{2m}; k - 1)]$ induced by the colouring c contains a monochromatic $(k - 1)$-term arithmetic progression.

This means that there are $k - 1$ blocks B_{i_j}, $1 \leq j \leq k - 1$, that are identically coloured by c and that are equally spaced between each other.

---

[3]The reason why we need $W(l^{2m}; k - 1)$ will become clear shortly.

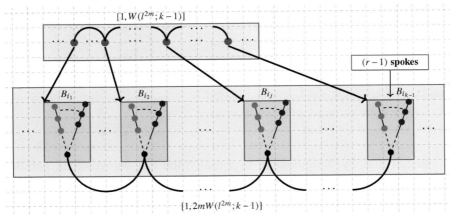

We observe that every block B_{i_j}, $1 \leq j \leq k-1$, contains $r-1$ colour-focused $(k-1)$-term arithmetic progressions. Note that from the definition of *colour-focused*, among those $r-1$ monochromatic arithmetic progressions there are no two that are coloured by the same colour. Also, since we assume that the colouring c does not contain a k-term monochromatic arithmetic progression, the colour of the focus is different than any of the colours already used.

We consider a new spoke that contains all foci and observe that there is a new spoke in each of the previously used colours with the same focus. The figure below depicts two new spokes with the common focus: one that contains all foci and one in a previously used colour.

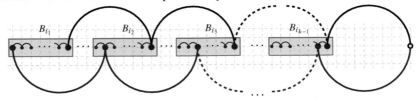

All foci form a spoke and there is a new spoke in each of the previously used colours. A set of r spokes appears!

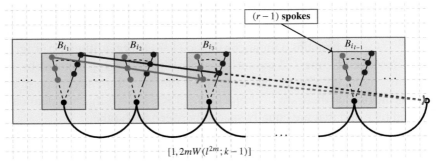

Therefore, for $M = 2mW(l^{2m}; k-1)$, every l-colouring of the interval $[1, M]$ contains a monochromatic k-term arithmetic progression or at least r colour–focused $(k-1)$-term arithmetic progressions. This completes the induction step in the proof of Claim 3.3.1. □

Where are you?

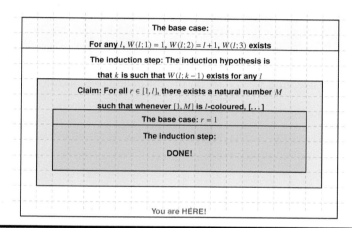

The base case:

For any l, $W(l; 1) = 1$, $W(l; 2) = l + 1$, $W(l; 3)$ **exists**

The induction step: The induction hypothesis is

that k is such that $W(l; k-1)$ exists for any l

Claim: For all $r \in [1, l]$, there exists a natural number M

such that whenever $[1, M]$ is l-coloured, [...]

The base case: $r = 1$

The induction step:

DONE!

You are HERE!

We complete the induction step in the proof of Theorem 3.1.1 by considering an l-colouring of the interval $[1, 2M]$, where M is the positive integer guaranteed by Claim 3.3.1 in the case $r = l$.

If we assume that the restriction of this colouring on interval $[1, M]$ does not contain a k-term monochromatic arithmetic progression then, by Claim 3.3.1, it must contain l-spokes with the focus in $[1, 2M]$. Since we have l spokes and l colours, by the pigeonhole principle, one of those spokes must be a k-monochromatic arithmetic progression.

Hence, we can take $n(l, k) = 2M$.

By the principle of mathematical induction, the number $n(l, k)$ exists for any $k, l \in \mathbb{N}$. □

3.4 VAN DER WAERDEN'S THEOREM: HOW FAR AND WHERE?

"Do not, however, confuse elementary with simple."

Aleksandr Yakovlevich Khinchin, a Soviet
mathematician
1894–1959

Recall that, for $k, l \in \mathbb{N}$, the van der Waerden number $W(l; k)$ is defined as the smallest positive integer N with the property that any l-colouring of the set $[1, N]$ contains a monochromatic k-term arithmetic progression.

In this section we reflect on the following two questions:

1. How big is $W(l; k)$?

2. If $\mathbb{N}$ is l-coloured can we be sure that a certain colour contains a k-term arithmetic progression?

3.4.1 Bounds for van der Waerden Numbers

Searching for the exact values of van der Waerden numbers and for improved lower and upper bounds has been an intriguing and challenging task for generations of mathematicians and computer scientists. For example, in 1974 Tom Brown reported that in 1971 the bound $W(2; 5) \geq 177$ was computed [13]. It took seven years more for the exact value $W(2; 5) = 178$ to be computed by Stevens and Shantaram [112].

In 2008, Michal Kouril, a Czech–American computer scientist, and Jerome L. Paul, an American mathematician and computer scientist, verified that $W(2; 6) = 1132$ [67]. In the conclusion of their article, Kouril and Paul stated:

Finding the value of van der Waerden numbers presents a challenging problem, since the underlying principle behind their computation is still unknown. Using preprocessing resulted in a significant reduction of the search space, which together with optimized SAT solvers and (eventually) hardware support in the form of Field Programmable Gate Arrays allowed for the computation of the sixth known van der Waerden number $W(2; 6) = 1132$. We have tested the same approach with $W(2; 7)$, and the preprocessing does not reduce the size of the search space sufficiently to be computable at this time.

Hence it should not be a surprise that the number of the, at this moment known, exact values of van der Waerden numbers is quite limited. See Table 3.1.

Table 3.1 The exact values and the best known lower bounds for van der Waerden numbers $W(l; k)$.

				k			
l	3	4	5	6	7	8	9
2	9	35	178	1132	> 3703	> 7584	> 27113
3	27	293	> 1209	> 8886	> 43855	> 238400	
4	76	> 1948	> 10437	> 90306	> 387967		
5	> 125	> 2254	> 24045	> 246956			
6	> 207	> 9778	> 56693	> 600486			

In 1952, Erdős and Rado introduced the van der Waerden's function $W : (l, k) \to W(l; k)$ and offered "what seems to be the first non-trivial lower estimate of $W(l; k)$."

Theorem 3.4.1 (Erdős and Rado [34]). *Let $k, l \in \mathbb{N} \setminus \{1\}$ and let m_0 be the largest integer such that $m_0^2 \leq 2kl^k$. Then there is an l-colouring of the interval $[1, m_0]$ that avoids monochromatic $(k + 1)$-term arithmetic progressions. In other words, $W(l; k + 1) > \sqrt{2kl^k}$.*

Proof. Suppose that $m \in \mathbb{N}$ is such that $m \geq W(l; k + 1)$, i.e. suppose that any l-colouring of $[1, m]$ contains a monochromatic $(k + 1)$-term arithmetic progression. Clearly, $m \geq k + 1$.

Let $r \in \mathbb{N}$ be such that $1 + kr \leq m < 1 + k(r + 1)$, i.e. let r be the largest integer for which there is a $(k + 1)$-term progression contained in $[1, m]$ with common difference r. It follows that, for any $d \in [1, r]$ and any $a \in [1, m - dk]$, the $(k + 1)$-term arithmetic progression $a, a + d, \ldots, a + kd$ is contained in $[1, m]$. Therefore the number of $(k + 1)$-term arithmetic progressions contained in $[1, m]$ equals to $M = \sum_{d=1}^{r} (m - kd) = mr - \frac{1}{2} kr(r + 1)$.

Recall that there are l^m distinct l-colourings of the interval $[1, m]$.

Say that $A = \{A_1, A_2, \ldots, A_M\}$ is the set of all $(k + 1)$-term arithmetic progressions contained in $[1, m]$. For each $i \in [1, M]$, let C_i be the set of all l-colourings of $[1, m]$ for which the arithmetic progression A_i is monochromatic. Since there are l distinct ways in which the progression A_i can be monochromatic and since, once the colour of A_i is fixed, there are l^{m-k-1} distinct ways to l-colour the remaining $m - k - 1$ elements of $[1, m]$, it follows that $|C_i| = l \cdot l^{m-k-1} = l^{m-k}$.

Observe that our assumption $m \geq W(l; k+1)$ implies, that for every l-colouring c of $[1, m]$, there is at least one $i \in [1, M]$ such that $c \in C_i$. Therefore

$$l^m = \left|\cup_{i=1}^M C_i\right| \leq \sum_{i=1}^M |C_i| = M l^{m-k},$$

which is equivalent to $l^k \leq M$.

This fact, together with the fact that $1 + kr \leq m < 1 + k(r+1)$, implies that

$$l^k \leq M = \frac{r}{2}(2m - kr - k) < \frac{(m-1)(m+1)}{2k} < \frac{m^2}{2k}.$$

Therefore $m > \sqrt{2kl^k}$. □

Fast forward to 2016, Jakub Kozik, a Polish mathematician, and Dmitry Shabanov, a Russian mathematician, proved that, for some absolute constant $c > 0$, $W(l; k) > cl^k$ [68].

In 1952, Erdős and Rado commented: "An upper estimate of [a van der Waerden number $W(l; k)$], at any rate one which is easily expressible explicitly in terms of the fundamental algebraic operations, seems to be beyond the reach of methods available at present" [34].

To fully appreciate Erdős and Rado's comment we need to go back to the 1920s and meet Wilhelm Ackermann, a German mathematician and logician, 1896–1962.

Part of Ackermann's mathematical legacy is the so-called *Ackermann function*.

We follow the notation and terminology used by Graham, Rothschild, and Spencer in [46] to obtain the Ackermann function in several steps.

We start with the function $f_1(x) = \text{DOUBLE}(x) = 2x$ and then define $f_2(x) = \text{EXPONENT}(x) = 2^x$.

Observe that $f_1^{(2)}(1) = f_1(f_1(1)) = f_1(2 \cdot 1) = 2 \cdot 2 = 2^2 = f_2(2)$, $f_1^{(3)}(1) = f_1(2^2) = 2 \cdot 2^2 = f_2(3)$, and in general $f_2(x) = f_1^{(x)}(1)$.

We continue and define

$$f_3(x) = \text{TOWER}(x) = \left.2^{2^{\cdot^{\cdot^{2}}}}\right\}x = \underbrace{f_2(f_2(f_2(\ldots f_2(1))))}_{x} = f_2^{(x)}(1).$$

Observe that $x \mapsto f_3(x) = \text{TOWER } (x)$ is growing very fast: $f_3(1) = 2$, $f_3(2) = 2^2 = 4$, $f_3(3) = 2^{2^2} = 16$, $f_3(4) = 2^{16} = 65,536$, and $f_3(5) = 2^{65,536}$.

Next, we define $f_4(x) = \text{WOW } (x) = f_3^{(x)}(1)$. For example, WOW $(3) = f_3(f_3(f_3(1))) = f_3(f_3(2)) = f_3(4) = 65,536$ and WOW $(4) = f_3(65,536)$.

Graham, Rothschild, and Spencer explain:

We call f_4 the WOW function. This fanciful description comes from trying to grasp the magnitude of $f_4(4)$ – a tower of twos of size $65,536$ – what can we say but "oh wow!" [46].

In general, for $i, x \in \mathbb{N}$, we define $f_{i+1}(x) = f_i^{(x)}(1)$.

It should be intuitively clear that, for any $i \geq 4$, each of the functions $x \mapsto f_i(x)$ is increasing at the rate that is incomprehensible to an ordinary human.

Next, imagine an infinite array in which every row and column is labeled by a natural number and in which the cell that is in the ith row and the xth column contains the integer $f_i(x)$. See Table 3.2.

Table 3.2 An infinite array with the (i, x)-entry of the form $f_i(x)$.

	1	2	3	$\ldots$	x	$\ldots$
1	$f_1(1)$	$f_1(2)$	$f_1(3)$	$\ldots$	$f_1(x)$	$\ldots$
2	$f_2(1)$	$f_2(2)$	$f_2(3)$	$\ldots$	$f_2(x)$	$\ldots$
3	$f_3(1)$	$f_3(2)$	$f_3(3)$	$\ldots$	$f_3(x)$	$\ldots$
$\vdots$	$\vdots$	$\vdots$	$\vdots$	$\vdots$	$\vdots$	$\vdots$
x	$f_x(1)$	$f_x(2)$	$f_x(3)$	$\ldots$	$f_x(x)$	$\ldots$
$\vdots$	$\vdots$	$\vdots$	$\vdots$	$\vdots$	$\vdots$	$\vdots$

We define the ACKERMANN function as the rule that associates to each natural number x the corresponding number on the diagonal in Table 3.2:

$$f_\omega(x) = \text{ACKERMANN } (x) = f_x(x).$$

For example, ACKERMANN $(4) = f_4(4) = \text{WOW } (4)$.

Since it captures such a rapid growth, the ACKERMANN function has been studied and applied in various mathematical settings. For our purpose, we mention the fact that, as demonstrated in [46, p. 64], the original van der Waerden's proof implies that, for $k \geq 10$, $W(2; k) \leq \text{ACKERMANN } (k)$.

Thus, Erdős and Rado's comment from 1952: "An upper estimate of $[W(l; k)]$ (...) seems to be beyond the reach of methods available at present."

A true breakthrough came in 1988 when Saharon Shelah, an Israeli mathematician, came up with a proof of the Hales–Jewett theorem that avoided the use of double induction [106]. Since the Hales–Jewett theorem implies van der Waerden's theorem (see Example 5.3.2), Shelah established that $W(2;k)$ was bounded from above by a WOW function.

It took more than 10 years for another breakthrough in the search for a tighter upper bound of $W(2;k)$. In the paper [42] published in 2001, Timothy Gowers, a British mathematician, proved that

$$W(2;k) < 2^{2^{2^{2^{2^{2^{k+9}}}}}}.$$

This settled Graham's conjecture from 1990 that $W(2;k) < \text{TOWER}(k)$.

The search for tighter upper bounds of van der Waerden numbers continues. It is possible that, to paraphrase Erdős and Rado, new breakthroughs are "beyond the reach of methods available at present." At the same time there is a feeling of certainty that those breakthroughs are coming. In Conlon's words:

It also feels like the difficulty of combinatorics has increased, partly because of this effect that a lot of strong people have come into it. And in order to do anything substantial, you really need to work quite hard and have some very substantial ideas. I think that both the breadth, and certainly the depth of combinatorics has changed since I started [64].

Conjecture 3.4.1 (Graham [48]). *($1,000) For all k, $W(2;k) < 2^{k^2}$.*

Ron Graham and Graham's number
To claim the prize, please contact Dr. Steve Butler.

3.4.2 Density of Sets and Arithmetic Progressions

van der Waerden's theorem guarantees that any finite colouring of the set of natural numbers contains monochromatic arithmetic progressions of any length. But, can we somehow check if our favourite colour contains long arithmetic progressions?

Before answering this question we need to briefly introduce the notion of the density of a set of natural numbers.

Definition 3.4.1 (Density of a Set). *Let A be a subset of the set of natural numbers* $\mathbb{N}$. *We define the upper density* $\overline{d}(A)$ *of the set A by* $\overline{d}(A) =$ $\limsup_{n\to\infty} \frac{|A\cap[1,n]|}{n}$. *Similarly,* $\underline{d}(A)$, *the lower density of A, is defined by* $\underline{d}(A) = \liminf_{n\to\infty} \frac{|A\cap[1,n]|}{n}$. *We say that A has density* $d(A)$ *if* $\underline{d}(A) = \overline{d}(A)$. *In that case,* $d(A) = \lim_{n\to\infty} \frac{|A\cap[1,n]|}{n}$.

Intuitively, the positive upper density of a set A, for a sequence of large values of n, represents the size ("percent") of $A \cap [1,n]$ relative to n, the size of $[1,n]$. Again intuitively, zero density means that, after some point, the elements of the set A are very sparse or do not appear at all.

Example 3.4.1. Let A be the set of all natural numbers divisible by 3. Then, for $n = 3k + i$, $k \in \mathbb{N}$ and $i \in \{0, 1, 2\}$,

$$\frac{|A\cap[1,n]|}{n} = \frac{k}{3k+i} = \frac{1}{3} - \frac{i}{3n} \to \frac{1}{3} \text{ when } n \to \infty.$$

Hence, not surprisingly, the density of the set A is $\frac{1}{3}$.

What is the density of the set of all powers of 2?

Let $B = \{2^k : k \in \mathbb{N}\}$ and let $n, k \in \mathbb{N}$ be such that $2^k \le n < 2^{k+1}$. Then

$$0 \le \frac{|B\cap[1,n]|}{n} = \frac{k}{n} \le \frac{\log_2 n}{n} \to 0 \text{ when } n \to \infty.$$

Hence, again not surprisingly, since the set of all powers of 2 is sparse, $d(B) = 0$.

The set of all powers of 2 is an example of an infinite set with its density equal to 0. Another infinite subset of the set of natural numbers with its density equal to 0 is the set of all prime numbers. One way to establish this fact is to use the celebrated Prime Number Theorem:

$$\frac{(\text{the number of primes}) \le x}{x} \approx \frac{1}{\log x} \text{ as } x \to \infty.$$

See, for example, [126].

Motivated by the fact that van der Waerden's proof gives "a very bad estimate" for $W(l; k)$, in 1936 Erdős and Turán considered the problem of finding, for a given n, the size $r_3(n)$ of the largest subset of $[1, n]$ that does not contain a three-term arithmetic progression [37]. Erdős and Turán concluded that "it is probable" that $\lim_{n\to\infty} \frac{r_3(n)}{n} = 0$.

Klaus Roth, a German-born British mathematician, 1925–2015, proved this as a fact in the early 1950s. A quarter of century later Endre Szemerédi, a Hungarian–American mathematician and computer scientist, proved that the same is true if $r_3(n)$ is replaced by $r_4(n)$, the size of the largest subset of $[1, n]$ that does not contain a four-term arithmetic progression [117].

In 1975, Szemerédi, in what Erdős called "a masterpiece of combinatorial ingenuity," settled the general case that, for any $k \geq 3$, $\lim_{n\to\infty} \frac{r_k(n)}{n} = 0$, where $r_k(n)$ is the size of the largest subset of $[1, n]$ that does not contain a k–term arithmetic progression, by proving the following statement:

Theorem 3.4.2 (Szemerédi's Theorem [118]). *For all $\varepsilon > 0$ and k, there exists $N = N(k, \varepsilon)$ such that if $n > N(k, \varepsilon)$ and $R \subseteq [0, n)$ with $|R| > \varepsilon n$ then R contains an arithmetic progression with k terms.*

Example 3.4.2. To see that Szemerédi's theorem implies van der Waerden's theorem observe that, for $k, l \in \mathbb{N}$, $\varepsilon \in \left(0, \frac{1}{l}\right)$, and $N = N(k, \varepsilon)$ guaranteed by Szemerédi's theorem, if we l-colour the set $[1, N]$, then at least one of the colour classes, say the colour class R, is of size at least $\frac{N}{l}$. Hence, $|R| \geq \frac{N}{l} > \varepsilon N$. By Szemerédi's theorem, this colour class contains an arithmetic progression of length k.

It is common to state Szemerédi's theorem in the following form:

Theorem 3.4.3 (Szemerédi's Theorem). *Any set of integers with positive upper density contains an arithmetic progression of any length.*

In Exercise 3.15, we demonstrate that Theorems 3.4.2 and 3.4.3 are equivalent to the statement that, for any $k \geq 3$, $\lim_{n\to\infty} \frac{r_k(n)}{n} = 0$.

The publication of Szemerédi's article [118] was not the end of the journey that Erdős and Turán started in 1936.

In 1977, Hillel Furstenberg, an American–Israeli mathematician, proved Szemerédi's theorem by using ergodic theory [39]. In the words of Bryna Kra, an American mathematician:

> This gave rise to the field of ergodic Ramsey theory, in which problems motivated by additive combinatorics are proven using ergodic theory. Ergodic Ramsey theory has since produced combinatorial results, some of which have yet to be obtained by other means [69].

Neither Szemerédi's proof nor Furstenberg's proof provided an improved upper bound for van der Waerden numbers, the problem that motivated Erdős and Turán's work in 1936. Only in 2001, Gowers's proof of Theorem 3.4.2, that generalized Roth's argument and used both Fourier analysis and combinatorics, did so [42]. As mentioned in the previous section, the bound obtained by Gowers is still the best known.

Observation 3.4.1. An infinite set of natural numbers with zero density may, or may not, contain long arithmetic progressions.

Consider, for example, the set B of all powers of 2. For $n, m \in \mathbb{N}$, $n < m$, from $2^n + 2^m = 2^n(1 + 2^{m-n})$ it follows that there are no $x, y, z \in B$ such that $x + y = 2z$, i.e. the set B does not contain a three-term arithmetic progression.

On the other hand, the set of all primes contains arbitrarily long arithmetic progressions, as was famously proved by Green and Tao [51].

Ben Green and Terence Tao

Szemerédi's theorem – another mighty oak planted by Erdős and his collaborators

I will tell you one story I think not so many people know about. Ron Graham and Erdős were great friends and Ron was a very generous, very kind man. So when he edited some volumes and needed a paper by Erdős, he just went ahead to write a paper – titled, *Some of my favourite problems and results* by Paul Erdős. This is a 1997 paper.

So these are Ron's words pretending to be Paul Erdős.

"Problems have always been an essential part of my mathematical life. A well-chosen problem can isolate an essential difficulty in a particular area, serving as a benchmark against which progress in this area can be measured. An innocent looking problem often gives no hint as to its true nature. It might be like a 'marshmallow' serving as a tasty tidbit supplying a few moments of fleeting enjoyment. Or it might be like an 'acorn' requiring deep and subtle new insights from which a mighty oak can develop."

So here Ron was describing Erdős's ability of formulating a problem. When Erdős saw this article in his name, he had a huge smile on his face about Ron putting words in his mouth: a marshmallow versus an acorn.

Fan Chung Graham
[72]

3.5 VAN DER WAERDEN'S THEOREM: SOME RELATED QUESTIONS

"Genres aren't closed boxes. Stuff flows back and forth across the borders all the time."

Simone de Beauvoir, a French philosopher, writer, social theorist, and feminist activist

1908–1986

It has been more than 100 years since Pierre Henry Joseph Baudet stated his fateful conjecture. Since then, the conjecture turned into van der Waerden's theorem which, together with van der Waerden's original proof, inspired generations of mathematicians to study the related questions. In this section, we mention several of those questions.

3.5.1 Mixed van der Waerden Numbers

We start by reminding the reader that the van der Waerden number $W(l; k)$ denotes the smallest of all positive integers n with the property that any l-colouring of the interval $[1, n]$ contains a monochromatic k-term arithmetic progression. We motivate expanding the notion of van der Waerden numbers by the following example:

Example 3.5.1. Consider a two-colouring $c : [1, 6] \to \{\bullet, \blacksquare\}$. Call $\bullet$ "colour 1" and call $\blacksquare$ "colour 2."

Observe that if there are at least two numbers coloured by $\bullet$, then there is a two-term $\bullet$-coloured progression.

If there is only one, or possibly none, number coloured by $\bullet$, then there is a three-term $\blacksquare$-coloured arithmetic progression.

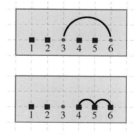

Hence, any two-colouring of $[1, 6]$ contains a monochromatic two-term arithmetic progression in colour 1 or a monochromatic three-term arithmetic progression in colour 2. Observe that 6 is the smallest number with this property. We write $W(2; 2, 3) = 6$, where the first number 2 denotes the number of colours, the second number 2 denotes the required length of an arithmetic progression monochromatic in colour 1 and the number 3 denotes the required length of an arithmetic progression monochromatic in colour 2.

We say that $W(2;2,3)$ is a *mixed van der Waerden number*.

In general, for $l, k_1, k_2, \ldots, k_l \in \mathbb{N}\backslash\{1\}$ and any fixed ordering $C_1, C_2, \ldots, C_l$ of the l colours, a mixed van der Waerden number $W(l; k_1, k_2, \ldots, k_l)$ is the least positive integer n with the property that any l-colouring of $[1, n]$ contains, for some $i \in [1, l]$, a k_i-term C_i-coloured arithmetic progression.

Theorem 3.5.1. *The mixed number $W(l; k_1, k_2, \ldots, k_l)$ exists for any choice of $l, k_1, k_2, \ldots, k_l \in \mathbb{N}\backslash\{1\}$ and any ordering of the available colours.*

Proof. Let $l, k_1, k_2, \ldots, k_l \in \mathbb{N}\backslash\{1\}$ and let $C_1, C_2, \ldots, C_l$ be an ordering of the colours. Let A be the set of all natural numbers n with the property that for any colouring $c : [1, n] \to \{C_1, \ldots, C_l\}$ there is, for some $i \in [1, l]$, a k_i-term C_i-coloured arithmetic progression. Observe that, to prove the theorem, it is enough to show that the set A is not empty.

Let $K = \max\{k_1, k_2, \ldots, k_l\}$ and let $N = W(l; K)$ be the corresponding van der Waerden number. Then, by van der Waerden's theorem, for any $c : [1, n] \to \{C_1, \ldots, C_l\}$ there is, for some $i \in [1, l]$, a K-term C_i-coloured arithmetic progression. But since $k_i \leq K$, this guarantees the existence of a k_i-term C_i-coloured arithmetic progression. Thus $N \in A$ and A is not an empty set. $\square$

Question 3.5.1. What are the known values of mixed van der Waerden numbers? If the exact value of a mixed van der Waerden number is not known, what are its lower and upper bounds?

To illustrate the fact that these questions have intrigued mathematicians and computer scientists over many years, here we mention several pioneering results from 1970 and a result from 2021 that has taken the Ramsey theory community by surprise.

In 1970, Vaclav Chvátal, a Czech–Canadian mathematician and computer scientist, by "using the computing facilities of the University of New Brunswick" found van der Waerden numbers $W(3;3) = 27$ and $W(2;4) = 35$, as well as the following mixed van der Waerden numbers: $W(2;3,4) = 18$, $W(2;3,5) = 22$, $W(2;3,6) = 32$, $W(2;3,7) = 46$, and $W(2;4,5) = 55$ [16].

The reader may have noticed that, for $k \in \{5,6,7\}$, $W(2;3,k) < k^2$. It turns out that this inequality also holds for all $k \in [8, 20]$. See, for example, [73].

For many years it was a widespread belief that $W(2;3,k) = O(k^2)$, i.e. that there is a positive number M and a natural number k_0 such that $W(2;3,k) \leq Mk^2$, for all $k \geq k_0$. For example, Green recalls: "I first heard the question of whether or not $W(2;3,k) = O(k^2)$ from Ron Graham in around 2004" [50].

Fast forward to 2021, to the surprise of many members of the Ramsey theory community [66], Green proved that $W(2;3,k)$ actually does not have a polynomial upper bound [50].

3.5.2 Canonical Form of van der Waerden's Theorem

We recall that van der Waerden's theorem guarantees that any finite colouring of positive integers contains long monochromatic arithmetic progression.

Example 3.5.2. Show that the two-colouring of positive integers

$$\underbrace{1}_{1}\ \underbrace{00}_{2}\ \underbrace{1111}_{4}\ \underbrace{0\cdots0}_{8}1\cdots\underbrace{1}_{16}0\cdots\underbrace{0}_{32}11\cdots$$

does not contain an infinite monochromatic arithmetic progression.

Solution. Note that, in this colouring, all integers in the interval $[16, 31]$ are coloured by the colour 1. In this colouring, for each natural number i, there is a monochromatic block of 2^i consecutive integers. Hence, there are monochromatic arithmetic progressions of any finite length.

Let A be an infinite arithmetic progression with the common difference d and let $n \in \mathbb{N}$ be such that $d < 2^n$.

Since $d < 2^n$, any interval of consecutive integers with at least 2^n elements must contain a term from A.

On the other hand, by definition of the colouring we have

$$\underbrace{11\cdots1}_{2^n}\underbrace{00\cdots0}_{2^{n+1}} \text{ or } \underbrace{00\cdots0}_{2^n}\underbrace{11\cdots1}_{2^{n+1}},$$

which implies that A is not monochromatic. $\qquad\qquad\square$

For an example of a two-colouring with bounded gaps, i.e. a colouring in which any interval of a certain prescribed length M meets both colours, that does not contain an infinite monochromatic progression, see Exercise 3.12.

Clearly if we colour each positive integer with a different colour, this colouring will not contain even a two-term monochromatic arithmetic progression. Therefore, in the statement of van der Waerden's theorem we cannot omit the condition that the colouring is finite.

Question 3.5.2. What exactly happens if we consider an infinite colouring of the set of positive integers?

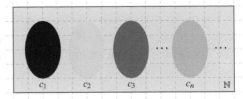

In 1980, Erdős and Graham [27] observed that Szemerédi's theorem guarantees the existence of $H(n)$, the smallest integer so that for any partition of $[1, H(n)]$ into any number of classes, there is always an n-term arithmetic monochromatic progression all of whose terms either belong to one class or all different classes.

This result became known as the canonical form of van der Waerden's theorem and it is commonly stated in the following way:

Theorem 3.5.2 (Canonical Form of van der Waerden's Theorem). *If f is an arbitrary colouring of the positive integers, then there are arbitrarily long monochromatic arithmetic progressions or arbitrarily long rainbow arithmetic progressions.*

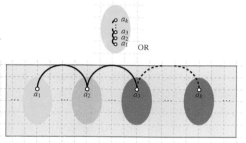

Here, *a rainbow arithmetic progression* means that no two terms of the arithmetic progression are of the same colour.

Proof. [4] Let $m > 1$ and let $\varepsilon \in \left(0, \frac{1}{m^2(m-1)}\right)$. Let the natural number n be large enough so that the interval $[0, n-1]$ contains more than $\frac{n^2}{4m}$ m-term arithmetic progressions (see Exercise 3.2). In addition, we chose n big enough so that Szemerédi's theorem applies, i.e. that any $A \subset [0, n-1]$ such that $|A| > \varepsilon n$ contains an m-term arithmetic progression.

Let $c : [0, n-1] \to C$ be a colouring.

If, for some $\alpha \in C$, $|c^{-1}(\alpha)| = |\{k \in [0, n-1] : c(k) = \alpha\}| > \varepsilon n$ then, by Szemerédi's theorem, there is an m-term arithmetic arithmetic progression contained in $c^{-1}(\alpha)$. In other words, there exists a c-monochromatic m-term arithmetic progression.

If, for all $\alpha \in C$, $|c^{-1}(\alpha)| < \varepsilon n$, by Exercise 3.5, there exists a c-rainbow m-term arithmetic progression. $\qquad\square$

Study of rainbow structures in *finite colourings* has lead to the development of the so-called Ramsey rainbow theory. For more details see, for example [38, 62].

---

[4]This proof, as well as Exercises 3.2 and 3.5, are based on the handwritten notes that Tom Brown shared with the author. Brown had taken those notes in 1980, during a demonstration of the proof by Ron Graham.

3.5.3 Polynomial van der Waerden's Theorem

We go back to van der Waerden's theorem, but now let us look at the common differences of the monochromatic arithmetic progressions guaranteed by the theorem.

Example 3.5.3. Let c be any l-colouring of the set of natural numbers. We define c', a new l-colouring, by $c'(n) = c(2n)$, for any $n \in \mathbb{N}$. Then, for any $k \in \mathbb{N}$, there is a c'-monochromatic arithmetic progression $a, a + d, a + 2d, \ldots, a + (k-1)d$. But, by definition, this implies that $2a, 2a + (2d), 2a + 2 \cdot (2d), \ldots, 2a + (k-1) \cdot (2d)$ is a k-term c-monochromatic arithmetic progression with the common difference equal to $2d$.

On the other hand, a two-colouring of $\mathbb{N}$ in which any two consecutive numbers are coloured differently, does not contain even a two-term monochromatic arithmetic progression with an *odd* common difference.

In short, for any finite colouring of $\mathbb{N}$ and any $k \in \mathbb{N} \backslash \{1\}$, there is a monochromatic k-term arithmetic progression with the common difference equal to $2n$, for some natural number n. At the same time, there is a two-colouring such that there is no a two-term monochromatic arithmetic progression with the common difference of the form $2n - 1$.

Question 3.5.3. Can we find other sets L with the property that any finite colouring of $\mathbb{N}$ must contain long monochromatic progressions with their common differences in L? Can we characterize the family $\mathcal{L}$ of such sets, i.e. can we establish the sufficient and necessary conditions that elements of $\mathcal{L}$ satisfy?

In another of defining moments in the development of Ramsey theory, in 1996, Vitaly Bergelson and Alexander Leibman proved polynomial extensions of van der Waerden's and Szemerédi's theorems [8]. A special case of their result determines a whole subfamily of $\mathcal{L}$.

Theorem 3.5.3 (Polynomial van der Waerden's Theorem – Special Case).

Let $l, k \in \mathbb{N}$ and let p be a non-zero polynomial with integer coefficients such that $p(0) = 0$. Then for any l-colouring of $\mathbb{Z}$ there are $a, d \in \mathbb{Z}$ such that the k-term arithmetic progression $a, a + p(d), a + 2p(d), \ldots, a + (k-1)p(d)$ is monochromatic.

Vitaly Bergelson and Alexander Leibman

In other words, if p is a non-zero polynomial with integer coefficients such that $p(0) = 0$ then $|p(\mathbb{N})| \cap \mathbb{N} = \{|p(n)| \in \mathbb{N} : n \in \mathbb{N}\} \in \mathcal{L}$.

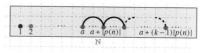

For example, if $p(x) = 2x$ then $p(0) = 0$ and, by Theorem 3.5.3, $p(\mathbb{N}) = 2\mathbb{N} \in \mathcal{L}$, exactly what we concluded in Example 3.5.3.

If $q(x) = 2x - 1$ then $q(0) = -1 \neq 0$, so the fact that $q(\mathbb{N}) = 2\mathbb{N} - 1 \notin \mathcal{L}$, the fact that we have already established, confirms that the condition $p(0) = 0$ in Theorem 3.5.3 is necessary.

Still, one should be careful here.

For example, let us consider polynomials $s(x) = x^2 - 2x$ and $r(x) = x^2 - 1$. Observe that, by the Polynomial van der Waerden's theorem, $|s(\mathbb{N})| \cap \mathbb{N} = \{1, 3, 8, 15, \ldots\} \in \mathcal{L}$. This implies that $r(\mathbb{N}) \cap \mathbb{N} = \{3, 8, 15, \ldots\} \in \mathcal{L}$ as well, regardless of the fact that $r(0) = -1 \neq 0$.

Not every element of $\mathcal{L}$ is obtained as, or from, a range of a polynomial. For example, $\{\lfloor \pi n \rfloor : n \in \mathbb{N}\} \in \mathcal{L}$. This was established in 1999 by Brown, Graham, and Landman in [14].

In their search for a characterization of the family $\mathcal{L}$, Brown, Graham, and Landman introduced the following terminology: For $l \in \mathbb{N}$, a set $L \subseteq \mathbb{N}$ is l-large if for any l-colouring of $\mathbb{N}$ and any $k \in \mathbb{N}\setminus\{1\}$, there is a monochromatic k-term arithmetic progression with the common difference in L. A set L is large if it is l-large for all l. In our notation, $\mathcal{L}$ is the family of all large sets.

At the time of writing this text, the following conjecture stated in 1999 by Brown, Graham, and Landman, was still unresolved:

Conjecture 3.5.1 (Two-large Conjecture [14]). *Every two-large set is large.*

Twenty years later: about the two-large conjecture

Mathematicians tend not to write of their failures. This is rather unfortunate as there are surely countless creative ideas that have never seen the light of day; I have long believed that a Journal of Failed Attempts should exist. My goal with this article is three-fold: (1) a chronicle of my battle with what I consider a particularly difficult conjecture; (2) to present my progress on this conjecture; and (3) to provide a roadmap to those who want to take on this challenging conjecture.

Aaron Robertson
[95]

3.6 EXERCISES

Exercise 3.1 (*Counting arithmetic progressions*). In this question we are interested in counting k-term arithmetic progressions in the given interval $[1, n]$.

1. Let $m \in \mathbb{N}$ be given and let $n = 3m + 1$.

    Let d be such that for some $a \in [1, n]$, the four-term arithmetic progression $a, a + d, a + 2d, a + 3d$ is contained in $[1, n]$.

    (a) Show that the number of four-term arithmetic progressions with the common difference d contained in $[1, n]$ is equal to $s(d) = n - 3d$.

    (b) Show that the maximum value of d is m.

    (c) Denote by $A_n(4)$ the number of four-term arithmetic progressions contained in the interval $[1, n]$. Show that $A_n(4) \leq \frac{n^2}{2 \cdot 3}$.

2. Let $k \in \mathbb{N}$ and let d be such that for some $a \in [1, n]$, the k-term arithmetic progression $a, a + d, a + 2d, \ldots, a + (k - 1)d$ is contained in $[1, n]$.

    (a) Show that the number of k-term arithmetic progressions with the common difference d contained in $[1, n]$ is equal to $s(d) = n - (k - 1)d$.

    (b) Show that the maximum value of d is $\lfloor \frac{n-1}{k-1} \rfloor$.

    (c) Denote by $A_n(k)$ the number of k-term arithmetic progressions contained in the interval $[1, n]$. Show that $A_n(k) \leq \frac{n^2}{2 \cdot (k-1)}$.

Exercise 3.2 (*Number of arithmetic progressions*). Prove that, for $m \in \mathbb{N}$ and a sufficiently large $n \in \mathbb{N}$, the number of m-term arithmetic progressions contained in $[0, n - 1]$ is greater than $\frac{n^2}{4m}$.

Exercise 3.3 (*van der Waerden's theorem and combinatorics of words*). Consider any infinite word $\mathcal{W}$ on the alphabet $\{A, B, C\}$. Prove that there is a two-letter word XY on the alphabet $\{A, B, C\}$ that will appear in $\mathcal{W}$ 1,000 times in such a way that 999 words that separate consecutive appearances of XY all have the same length.

Exercise 3.4 (*Counting monochromatic arithmetic progressions*).

1. Let $c : \mathbb{N} \rightarrow \{1, 2\}$ be a two-colouring of $\mathbb{N}$ and let $k \in \mathbb{N}$. Prove that there exists at least one colour $i \in \{1, 2\}$ such that there is an infinite number of monochromatic k-term arithmetic progressions in that colour.

2. If in the first part of this exercise you replace "a two-colouring" by "an r-colouring, $r \geq 2$," would the statement still be true? In other words, is it true that, for any finite colouring of the natural numbers and any (fixed) $k \in \mathbb{N}$, there is always an infinite number of monochromatic k-term arithmetic progressions in a single colour? Why yes, or why not?

3. If in the first part of this exercise you replace "an infinite number of monochromatic k-term arithmetic progressions" by "an infinite monochromatic progression," would the statement still be true? In other words, is it true that, for any two-colouring of the natural numbers, there is always an infinite monochromatic arithmetic progressions? Why yes, or why not?

Exercise 3.5 (*Colour class size and rainbow arithmetic progressions*). Let m be an integer greater than 1 and let $\varepsilon \in \left(0, \frac{1}{m^2(m-1)}\right)$. If the natural number n is large enough so that the interval $[0, n-1]$ contains more than $\frac{n^2}{4m}$ m-term arithmetic progressions (as established in Exercise 3.2), prove that the following is true:

Let $c : [0, n-1] \to C$ be any colouring such that, for each $\alpha \in C$, $|c^{-1}(\alpha)| = |\{k \in [0, n-1] : c(k) = \alpha\}| \leq \varepsilon n$. Then there exists an m-term c-rainbow arithmetic progression. In other words, there is a set $P = \{a, a+d, \ldots, a+(m-1)d\} \subseteq [0, n-1]$ such that, for $i, j \in [0, m-1]$, $i \neq j$ implies that $c(a+id) \neq c(a+jd)$.

Exercise 3.6 (*Special common difference*). Let $\chi : \mathbb{N} \to \{0, 1\}$ be a two-colouring of positive integers and let $k \in \mathbb{N}$. Use van der Waerden's theorem to prove that there is a χ-monochromatic k-term arithmetic progression with a common difference that is divisible by 3.

Exercise 3.7 (*The difference and all terms of the same colour*). Prove that for any $k, l \in \mathbb{N}$ there is $S(k, l) \in \mathbb{N}$ such that any k-colouring of the set of positive integers $[1, S(k, l)]$ contains an arithmetic progression of length l with the property that its common difference and all its terms are of the same colour.

Exercise 3.8 (*Syndetic sets*). This question is about so-called syndetic sets.

A syndetic set is any set that can be represented as an increasing sequence of positive integers with bounded gaps, i.e. as a sequence $a_1 < a_2 < a_3 < \ldots$ such that there is $M > 0$ for which $a_{i+1} - a_i \leq M$, for all $i \in \mathbb{N}$.

1. Which of the following two sets is syndetic:

    (a) The set of all (positive) powers of 2.

    (b) The set of all positive integers that are divisible by 3 or by 5.

2. Prove that if $\mathbb{N} = A \cup B$, then either A contains arbitrarily long intervals or B is syndetic.

3. Which of the following two sets contains long arithmetic progressions:

    (a) The set of all (positive) powers of 2.

    (b) The set of all positive integers that are divisible by 3 or by 5.

4. Prove that every syndetic set contains long arithmetic progressions.

5. If the set of positive integers is partitioned into two classes, then at least one of the following holds:

    (a) One class contains arbitrarily long intervals of consecutive integers.

    (b) Both classes contain arithmetic progressions of arbitrary length.

Exercise 3.9 (*Syndetic sets – an application*). You programmed your computer to do the following.

At each step $i \geq 1$, your algorithm creates independently two sets of positive integers, $A(i)$ and $B(i)$, in the following way:

1. Randomly choose positive integers, call them $a(1)$ and $b(1)$. Set $A(1) = \{a(1)\}$ and $B(1) = \{b(1)\}$.

2. For $i > 1$, randomly choose positive integers $a(i)$ and $b(i)$ so that $0 < a(i) - \max A(i-1) \leq 10^3$ and $0 < b(i) - \max B(i-1) \leq 10^6$. Set $A(i) = A(i-1) \cup \{a(i)\}$ and $B(i) = B(i-1) \cup \{b(i)\}$.

After the algorithm produces $A(i)$ and $B(i)$, the program plots the set of points $P(i) = A(i) \times B(i) = \{(x, y) : x \in A(i), y \in B(i)\}$ and checks if there is a line $\ell(i)$ with a positive slope such that $\ell(i) \cap P(i)$ contains a set of at least 10^9 equidistant points. In other words, you are looking for a set $T(i) \subseteq \ell(i) \cap P(i)$ such that $|T(i)| \geq 10^9$ and that for any two consecutive points $U, V \in T(i)$, $|\overline{UV}| = s$, for some (fixed) $s > 0$.

The program stops if it finds the line $\ell(i)$. If such a line does not exist, the program continues with step $i + 1$.

Is there a possibility that your program runs forever? Why yes or why not?

Exercise 3.10 (*van der Waerden game*). Two players, Maker and Breaker, play the following game on a one-way endless strip of empty 1×1 cells: Maker and Breaker take turns. On their turn, Maker puts a triangle into any empty cell, and they wins if they forms a three-term arithmetic progression of triangles. On their turn, Breaker puts 1,000 circles into any 1,000 (not necessarily adjacent) distinct empty cells, and they wins if they prevents Maker from forming a three-term arithmetic progressions of triangles.

1. Show that Maker can win if they goes first.
2. What would happen if the players switch the order who starts the game?
3. What would happen if "three-term arithmetic progression of triangles" is replaced by "k-term arithmetic progression of triangles?"

Exercise 3.11 (*Arithmetic progressions and a sequence generated by an irrational number*). Let α be a positive irrational number. Then the set $S_\alpha = \{[k\alpha] : k \in \mathbb{N}\}$ contains long arithmetic progressions. *Note:* $[x]$ denotes the nearest integer function: $[x] = k \in \mathbb{Z}$, where $|k - x| \leq 0.5$.

Exercise 3.12 (*Infinite arithmetic progressions and a sequence generated by an irrational number*). Let $\alpha > 1$ be an irrational number. Then the set $S_\alpha = \{[k\alpha] : k \in \mathbb{N}\}$ does not contain an infinite arithmetic progression. *Note:* Use that fact that, if $\beta = \frac{\alpha}{\alpha-1}$, then $S_\alpha \cup S_\beta = \mathbb{N}$ and $S_\alpha \cap S_\beta = \emptyset$. For a proof of this fact see, for example, [6].

Exercise 3.13 (*Mixed van der Waerden number*). Prove that $W(3; 2, 3, 3) \leq 18$.

Exercise 3.14 (*Mixed van der Waerden number*). Show that if $k \equiv \pm 1 \pmod 6$ then $W(3; k, 2, 2) = 3k$.

Exercise 3.15 (*Three equivalent statements of Szemerédi's theorem*). Three common ways of stating Szemerédi's theorem are:
A: Let $k \geq 3$. For $n \in \mathbb{N}$, let $r_k(n)$ be the size of the largest subset of $[1, n]$ that does not contain a k-term arithmetic progression. Then $\lim_{n \to \infty} \frac{r_k(n)}{n} = 0$.
B: For all $\varepsilon > 0$ and $k \in \mathbb{N}$, there exists $N = N(k, \varepsilon)$ such that if $n > N(k, \varepsilon)$ and $R \subseteq [0, n)$, with $|R| > \varepsilon n$, then R contains an arithmetic progression with k terms.
C: Any set of integers with positive upper density contains an arithmetic progression of any length.
    Prove that: $\mathbf{A} \Rightarrow \mathbf{B} \Rightarrow \mathbf{C} \Rightarrow \mathbf{A}$, i.e. prove that the three statements are equivalent to each other.

Exercise 3.16 (*Syndetic sets and Szemerédi's theorem*). Use Szemerédi's theorem to prove that any syndetic set contains long arithmetic progressions.

CHAPTER 4

Schur's Theorem and Rado's Theorem

Schur's THEOREM is one of the earliest results in combinatorial number theory, a mathematical field that is interested in questions that lie between number theory and combinatorics. As we will see, Schur's theorem, a combinatorial fact, is used to prove seemingly unrelated number theory facts.

Rado's theorem is a generalization of Schur's theorem and another of the milestones in the development of Ramsey theory.

We briefly talk about Schur and Rado, two remarkable people whose mathematical legacies still live on today. We will present a few examples of Schur's and Rado's work.

As before, the end of the chapter includes exercises related to the chapter's topic.

4.1 ISSAI SCHUR

"To live without hope is to cease to live."

Fyodor Mikhailovich Dostoyevsky, a Russian novelist

1821–1881

4.1.1 Who Was Issai Schur?

Issai Schur, was one of the great mathematicians of the first part of the twentieth century, a professor at the University of Berlin, Germany, and a member of the Prussian Academy, who died in exile in 1941.

DOI: 10.1201/9781003286370-4

A list of mathematical objects, notions, and techniques named after Issai Schur:

Frobenius–Schur indicator	Herz–Schur multiplier
Jordan–Schur theorem	Lehmer–Schur algorithm
Schur algebras	Schur complement
Schur complement method	Schur decomposition
Schur index	Schur function
Schur indicator	Schur multiplier
Schur orthogonality relations	Schur polynomial
Schur's inequality	Schur product
Schur–convex function	Schur–Horn theorem
Schur–Weyl duality	Schur–Zassenhaus theorem
Schur's lemma	Schur test
Schur's lower bound	Schur's property

Schur's theorems in Ramsey theory, differential geometry, linear algebra, analysis

Birth and Death. Issai Schur was born on January 10, 1875, in Mogilev, Russian Empire (now Belarus), and died on January 10, 1941, in Tel Aviv, Palestine (now Israel).

World in 1875

World in 1941

Electric dental drill is patented.
Georges Bizet's opera *Carmen* premieres.
The US Congress passes the Civil Rights Act.
Alexander Graham Bell makes the first voice transmission.
A rebellion against the Ottoman rule broke out in Bosnia and Herzegovina.
The first woman licensed to practise medicine in Canada.
The Metre Convention Treaty signed.

War!

Issai's Family. Issai Schur's parents were Golde Landau and Moses Schur, a merchant. In 1906, he married Regina Malka Frumkin, a medical doctor. They had two children, Georg and Hilde.

Besides being one of the dominant figures in mathematical research of his time, Schur was also known as an outstanding lecturer. In December 2000, at the Schur Memoriam Workshop held at the Weizmann Institute, Walter Ledermann, a German and British mathematician, 1911–2009, remembered:

> I attended many courses. But Schur's lectures were for me the most impressive and inspiring ones. It seemed to me that they were perfect both in content and in form. Schur was a superb lecturer. He spoke slowly and clearly and his writing on the blackboard was very legible. All his courses were carefully structured into chapters and sections, each bearing a number and an appropriate heading. His lectures were meticulously prepared. It is known that he had very full lecture notes, written on loose sheets which he carried in the breast pocket of his jacket. But I can remember only one occasion when he consulted his notes: during one of the lectures on invariants he wrote down a list of invariants of a certain quintic polynomial. He furtively pulled out a sheet of paper from his pocket in order to check whether he had remembered the rather complicated formulae correctly (he had!). He never got stuck in his lecture or failed to remember what he had said in the previous lecture [75].

In April 1933, all Jewish government officials in Germany were dismissed, a boycott of Jewish businesses was decreed, and antisemitic legislation had begun. These events catastrophically affected the life of every German Jew, including Schur.

In 1986, at a symposium dedicated to the memory of Issai Schur at Tel Aviv University, Menahem Max Schiffer, a German–born American mathematician, 1911–1997, gave a lecture in which he talked about "the human dimensions of the man and the difficult and even tragic circumstances under which some of his work was accomplished" [103]. In his lecture, Schiffer provided several first-hand testimonials about Schur's life between 1933 and 1941.

Here is one of those testimonials that also serves as a reminder to everyone that even small acts of kindness may have a profound effect on those in need:

> Schur told me [in Palestine] that the only person at the Mathematical Institute in Berlin who was kind to him was [Helmut] Grunsky [a German mathematician, 1904–1986], then a young lecturer. Long after

the war, I talked to Grunsky about that remark and he literally started to cry: "You know what I did? I sent him a postcard to congratulate him on his sixtieth birthday. I admired him so much and was very respectful in that card. How lonely he must have been to remember such a small thing [103]."

Schur left Germany for Palestine in 1939, broken in mind and body.

"In 1941 he celebrated his last birthday. His son Georg had volunteered for the British army and was already in uniform on that occasion. The excitement was probably too much for Schur. Soon after, he had his final heart attack and we lost him" [103].

Erdős and Schur:

In an interview with Gerald Lee Alexanderson, an American mathematician, 1933–2020, Erdős recalled:

Issai Schur was one of the first foreigners with whom I corresponded. I proved that there is a prime of the form $4k + 1$ and $4k + 3$ between n and $2n$.[1] Schur was very impressed and he wrote me a nice letter in German. It pleased my parents.

This was before the Nazis, in 1932. Later I visited Schur. I saw him in 1936 and I saw him in 1938 for the last time. My mother and I visited Mrs. Schur in Israel just before her death. I visited her every year; but my mother first came to Israel in 1964–65, and when we were in Tel Aviv we visited her [2].

4.1.2   Schur's Work: Two Examples

4.1.2.1   Schur Complement

In 1917, Schur established the necessary and sufficient conditions for all roots of the polynomial $f(x) = a_n x^n + a_{n-1} x^{n-1} + \ldots + a_1 x + a_0$, $a_i \in \mathbb{C}$, to lie inside the unit circle [105]. In the process, Schur proved the following lemma:

Lemma 4.1.1 (Schur Determinant Lemma [127])**.** *Let P, Q, R, S denote four $n \times n$ matrices and suppose that P and R commute. Then the determinant $det(M)$ of the $2n \times 2n$ matrix $M = \begin{bmatrix} P & Q \\ R & S \end{bmatrix}$ is equal to the determinant of the matrix $PS - RQ$.*

---

[1]This was Erdős's first published paper, [25]. He was 18 years old at the time.

Proof. Let I and 0 denote the identity and the zero matrices respectively.

If P is a non-singular matrix then

$$\begin{bmatrix} P^{-1} & 0 \\ -RP^{-1} & I \end{bmatrix}\begin{bmatrix} P & Q \\ R & S \end{bmatrix} = \begin{bmatrix} I & P^{-1}Q \\ 0 & S - RP^{-1}Q \end{bmatrix}.$$

Taking determinants we obtain $\det(P^{-1}) \cdot \det(M) = \det(S - RP^{-1}Q)$ and so $\det(M) = \det(P)\det(S - RP^{-1}Q) = \det(PS - PRP^{-1}Q) = \det(PS - RQ)$.

If, however, $\det(P) = 0$, we replace matrix M with the matrix $M_\varepsilon = \begin{bmatrix} P + \varepsilon I & Q \\ R & S \end{bmatrix}$, for some $\varepsilon > 0$. The matrices R and $P + \varepsilon I$ commute. For a sufficiently small $\varepsilon > 0$, the determinant of $P + \varepsilon I$ is not equal to 0 and so $\det(M_\varepsilon) = \det((P + \varepsilon I)S - RQ)$. Letting ε converge to 0 yields the desired result. □

Seventy years after Schur established Lemma 4.1.1, in 1968, Emilie Haynsworth, an American mathematician, 1916–1985, introduced the term *Schur complement*.

Definition 4.1.1 (Schur Complement). *If P is a non-singular matrix then the Schur complement of the block P of the matrix $M = \begin{bmatrix} P & Q \\ R & S \end{bmatrix}$ is the matrix $S - RP^{-1}Q$.*

The Schur complement of the block P of the matrix M is commonly denoted by $M/P = S - RP^{-1}Q$. Observe that in the case of a non-singular matrix P, the conclusion of the lemma may be written as $\det(M)/\det(P) = \det(S - RP^{-1}Q)$.

For the properties of the Schur complement and its many applications in modern mathematics, see [127], a book written with the goal "to expose the Schur complement as a rich and basic tool in mathematical research and applications and to discuss many significant results that illustrate its power and fertility."

4.1.2.2 Schur and Ramsey Theory

In 1637, Pierre de Fermat, a French mathematician, 1607–1665, scribbled into the margins of his copy of *Arithmetica* by Diophantus, that

> It is impossible for a cube to be the sum of two cubes, a fourth power to be the sum of two fourth powers, or in general for any number that is a power greater than the second to be the sum of two like powers. I have discovered a truly marvellous demonstration of this proposition that this margin is too narrow to contain.

The margin note became known as Fermat's Last Theorem. It was proved by Andrew Wiles, an English mathematician, in 1995.

In an effort to shed more light on Fermat's Last Theorem, in 1916 Schur proved the following:

Theorem 4.1.1 ([104]). *Let $n > 1$. There exists an integer $M = M(n)$ such that, for all primes $p > M$, the congruence $x^n + y^n \equiv z^n \pmod{p}$ has a solution in the integers, such that p does not divide xyz.*

In his proof, Schur used the following *einfachen Hilfssatz*, i.e. *an easy lemma*, that we now call Schur's theorem:

Theorem 4.1.2 (Schur's Theorem [104]). *If you distribute the numbers $1, 2, \ldots, N$ on m rows then, once $N > m!e$, at least one row contains two numbers whose difference is also contained in the same row.*

If we think about each row as a cell of a partition of $[1, N]$ into m parts, then Theorem 4.1.2 states there is one part (a colour!) that contains integers x, y, z such that $y = z - x$. In other words, any m-colouring of $[1, N]$ contains a monochromatic solution of the equation $x + y = z$.

It turned out that Schur's theorem was one of those results that, a century later, Spencer characterized as "different theorems [that] have common methodologies and this common theme to them" [70]. Those common methodologies and the common theme were the starting point for building Ramsey theory as we know it today.

Moreover, in 1933, Schur's student Richard Rado in his doctoral thesis *Studien zur Kombinatorik* completely solved the following problem:

Let $AX = 0$ be a system of linear equations, where A is a matrix with integer entries. Under which conditions, for every r-colouring of the set of positive integers, does the system have a monochromatic solution?

Schur's theorem and van der Waerden's theorem, in the case of three-term arithmetic progressions, are two special cases of Rado's result. To see this we take $X = \begin{bmatrix} x \\ y \\ z \end{bmatrix}$ and observe that, for $A = \begin{bmatrix} 1 & 1 & -1 \end{bmatrix}$, the equation $AX = 0$ becomes $x + y - z = 0$ and that, for $A = \begin{bmatrix} 1 & 1 & -2 \end{bmatrix}$, the equation $AX = 0$ becomes $x + y - 2z = 0$.

4.2   SCHUR'S THEOREM

"My methods are really methods of working and thinking; this is why they have crept in everywhere anonymously."

Emmy Noether

In this section we reformulate and prove Theorem 4.1.2.

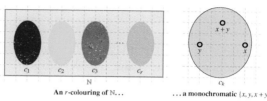

An r-colouring of $\mathbb{N}$...   ...a monochromatic $\{x, y, x+y\}$

Definition 4.2.1 (Schur Triple). *A triple (x, y, z) that satisfies the equation $x + y = z$ is called a Schur triple.*

We observe that, similar to the case of monochromatic arithmetic progressions, we do not need to colour the whole set of natural numbers to be guaranteed a monochromatic Schur triple. For the purpose of our proof, we slightly reformulate the statement of Schur's *einfachen Hilfssatz*.

Theorem 4.2.1 (Schur's Theorem). *Let $r \in \mathbb{N}$. There is a natural number M such that any r-colouring of the segment of positive integers $[1, M]$ contains a monochromatic Schur triple.*

Proof. Let $c : \mathbb{N} \to [1, r]$ and let M be the Ramsey number $R(\underbrace{3, 3, \ldots, 3}_{r})$.

Denote the vertices of the complete graph K_M by $1, 2, \ldots, M$. For any $a, b \in [1, M]$, we colour the edge $\{a, b\}$ in K_M by $c(|a - b|)$. Observe that, for this edge r-colouring of K_M, all we need is the restriction of the r-colouring c on the interval $[1, M - 1]$.

Because of our choice of the number M, there is a monochromatic triangle with the vertices $i < j < k$.

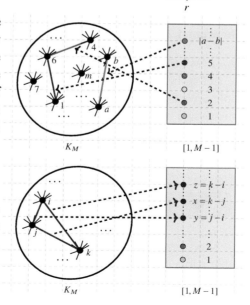

We take $x = k - j$, $y = j - i$, and $z = k - i$ and observe that $c(x) = c(y) = c(z)$ and $x + y = (k - j) + (j - i) = k - i = z$. □

Definition 4.2.2 (Schur Number). *The least M with the property that any r-colouring of $[1, M]$ contains a monochromatic Schur triple is called a* Schur number *and is denoted by $s(r)$.*

Example 4.2.1. What is $s(2)$?
Can we two-colour, say in red and blue, the interval of positive integers $[1, 4]$ and avoid monochromatic Schur triples? Note that $(1, 1, 2)$ and $(2, 2, 4)$ are Schur triples. Hence, $s(2) > 4$.
Can we two-colour the interval of positive integers $[1, 5]$ and avoid monochromatic Schur triples?
Therefore, $s(2) = 5$.

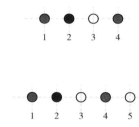

By now, the reader should not be surprised that there are only a few known Schur numbers at this moment: $s(1) = 2, s(2) = 5, s(3) = 14, s(4) = 45$, and $s(5) = 161$.

Observe that from the proof of Theorem 4.2.1 and Exercise 2.17 it follows that $s(r) \leq 3r! - 1$. Recall, see Theorem 4.1.2, that Schur established that $s(r) \leq er! + 1$.

Here is a lower bound for $s(r)$ established by Schur in 1916.

Theorem 4.2.2. *For $r \in \mathbb{N}$, $s(r) \geq \frac{3^r + 1}{2}$.*

Proof. We first prove that $s(r + 1) \geq 3s(r) - 1$.

Let χ be an r-colouring of $[1, s(r) - 1]$ with no monochromatic Schur triples. We extend the colouring χ to an $(r + 1)$-colouring χ' of $[1, 3s(r) - 2]$ as is shown in the figure below.

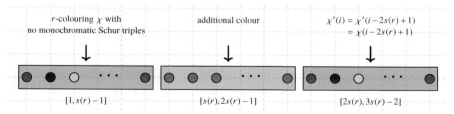

Now, observe that neither $[s(r), 2s(r) - 1]$ nor $[2s(r), 3s(r) - 2]$ contains a Schur triple. By definition, $[1, s(r) - 1]$ does not contain a χ'-monochromatic Schur triple. If $x \in [1, s(r) - 1]$ and $z \in [2s(r), 3s(r) - 2]$ then $z - x \in [s(r) +$

$1, 3(s(r) - 1)]$. We distinguish two cases. If $z - x \in [s(r) + 1, 2s(r) - 1]$ then the Schur triple $(x, z - x, z)$ is not χ'-monochromatic. Let $z - x \in [2s(r), 3(s(r) - 1)]$. This implies that the Schur triple $(x, z - x - 2s(r) + 1, z - 2s(r) + 1)$ is contained in the set $[1, s(r) - 1]$ and therefore is not χ'-monochromatic. It follows that the Schur triple $(x, z - x, z)$ is not χ'-monochromatic, Hence $s(r + 1) > 3s(r) - 2$ or, what is the same, $s(r + 1) \geq 3s(r) - 1$.

Recall that $s(1) = 2 = \frac{3^1 + 1}{2}$. For $r \in \mathbb{N}$ such that $s(r - 1) \geq \frac{3^{r-1} + 1}{2}$ we obtain,

$$s(r) \geq 3s(r - 1) - 1 \geq 3 \cdot \frac{3^{r-1} + 1}{2} - 1 = \frac{3^r + 1}{2}.$$

□

In the late 1950s, Leo Moser, a Canadian mathematician, 1921–1970, in his lecture entitled *Combinatorial Number Theory*, used Schur's theorem as the first example of "many interesting questions that lie between number theory and combinatorial analysis" [84].

Moser also commented that Schur's theorem was "related in a surprising way to Fermat's Last Theorem." In the rest of this section we explore this relationship, i.e. we prove Theorem 4.1.1.

Observation 4.2.1. For any odd prime p, the multiplicative group $\mathbb{Z}_p^* = \mathbb{Z}/p\mathbb{Z} = \{1, 2, \ldots, p - 1\}$ is *cyclic*, i.e. there is $q \in \{1, \ldots, p - 1\}$ such that $\mathbb{Z}_p^* = \{q, q^2, \ldots, q^{p-1}\}$.

For example, if we take $p = 5$ then $\mathbb{Z}_5^* = \{1, 2, 3, 4\}$ and the multiplication is given by

·	1	2	3	4
1	1	2	3	4
2	2	4	1	3
3	3	1	4	2
4	4	3	2	1

Observe that $\mathbb{Z}_5^* = \{2, 2^2, 2^3, 2^4\} = \{2, 4, 3, 1\}$.

Proof of Theorem 4.1.1.

Let $n \in \mathbb{N}\setminus\{1\}$ and let $p > s(n)$ be a prime. Choose $q \in [2, p - 1]$ so that $\mathbb{Z}_p^* = \{q, q^2, \ldots, q^{p-1}\}$.

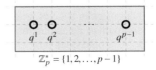

$$\mathbb{Z}_p^* = \{1, 2, \ldots, p - 1\}$$

Let χ be an n-colouring of $\mathbb{Z}_p^*$ with the property that

$\chi(q^u) = \chi(q^v)$
if and only if
$u \equiv v \pmod{n}$.

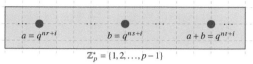

Since $p > s(n)$, there is a monochromatic Schur triple $(q^{nr+i}, q^{ns+i}, q^{nt+i})$.

Observe that $q^{nr+i} + q^{ns+i} = q^{nt+i}$ implies $q^i(q^{nr} + q^{ns} - q^{nt}) \equiv 0$ (mod p). Since $1 \le i \le n < s(n) < p$ and $\{q, q^2, \dots, q^{p-1}\} = \mathbb{Z}_p^*$, it follows that $p \nmid q^i$. Therefore $p | (q^{nr} + q^{ns} - q^{nt})$, or, what is the same, $q^{nr} + q^{ns} - q^{nt} \equiv 0$ (mod p).

By taking $x = q^r$, $y = q^s$, and $z = q^t$ we obtain $x^n + y^n \equiv z^n$ (mod p). Observe that $p \nmid xyz = q^{r+s+t}$. $\qquad\square$

Schur's theorem and beyond:

   Over the last century, there has been a steady stream of generalizations of Schur's theorem and questions and results inspired by the theorem. See, for example, [24, 73, 96, 102].

   Early in his career Julian Sahasrabudhe, a Canadian mathematician, proved a variation on the theme of Schur's theorem [101]. In his conversation with a group of undergraduate students he reflected on that experience:

"You've seen Schur's theorem; If you finitely colour the integers then you can find positive x, y and $x + y$ all in the same colour. Great. Using this, one can also prove a multiplicative version of Schur's theorem: you have x, y and then $x \cdot y$ all in the same colour.

One of my first big results was to show actually that there is an exponential version of this as well: you can find integers a and b, both say bigger than one, so that a, b and a^b are all in the same colour. It sounds simple when you put it next to Schur's results, which has been known for over a hundred years. But this was genuine new mathematics and pretty surprising that the exponential behaves the same.

Tom Brown and I discussed the version for two colours. That got me interested in the problem; I then managed to prove it for three and four colours. I spent quite a lot of time trying to disprove it for more colours because it seemed so ridiculous that the exponential would behave like $x + y$ and like $x \cdot y$. I spent quite a bit of time trying to find a colouring that would not contain a, b, and a^b all in the same colour. I had various partial colourings and was trying to stitch them together, with no success. And then I was like, OK, maybe I should try to prove that the claim is actually true.

That's a happy ending story" [71].

4.3 RICHARD RADO

"There are almost as many types of mathematicians as there are types of human beings. Among them are technicians, there are artists, there are poets, there are dreamers, men of affairs, and many more."

Richard Rado

4.3.1 Who Was Richard Rado?

Richard Rado was a mathematician who earned two PhDs: in 1933 from the University of Berlin for the thesis *Studien zur Kombinatorik* supervised by Issai Schur, and in 1935 from the University of Cambridge for the thesis *Linear transformations of sequences* supervised by Godfrey Harold Hardy, an English mathematician, 1877–1947. In 1986, Rado was awarded an Honorary Doctor of Mathematics Degree from the University of Waterloo.

Richard Rado's mathematical interests included:

Convergence of sequences and series	Inequalities	Ramsey theory
Geometry and measure theory	Number theory	Graph theory

Birth and Death. Richard Rado was born on April 28, 1906, in Berlin, Germany, and died on December 23, 1989, in Henley–on–Thames, Oxfordshire, England.

World in 1906
The first woman elected to American Society of Civil Engineers.
Axel Thue uses the Thue–Morse sequence to found the study of combinatorics on words.
Albert Einstein introduces his Theory of Relativity.
Mahātmā Gandhi coins the term "Satyagraha."
Andrey Markov produces his first theories on Markov chain processes.
SOS adopted as warning signal.
Lee de Forest patents a three-diode amplification valve.

World in 1989
Fall of the Berlin wall.
The first graphical user interface for Maple was developed.
Beginning of the dismantlement of the apartheid system in South Africa.
Leslie Lamport turns over development of LaTeX to Frank Mittelbach.
Warner Bros. release *Batman* with Michael Keaton and Jack Nicholson.
Tim Berners–Lee introduces a system that would become the World Wide Web.
Geoffrey Exoo establishes that $43 \leq R(5,5) \leq 49$.

Richard's Family. Richard was the second son of Anna Abrahamson and Leopold Rado. As a young man he had to choose between being a concert pianist or a mathematician. He chose to become a mathematician in the belief that he could continue with music as a hobby, but that he could never treat mathematics in that way. In 1933, Rado married Luise Zadek, whom he had met when he needed a partner to play piano duets. They had one son, Peter.

In 1991, Claude Ambrose Rogers, an English mathematician, 1920–2005, remembered:

> Rado and his wife had a double partnership: she went with him to mathematical conferences and meetings and kept contact with his mathematical friends, he was an accomplished pianist and she was a singer of professional standard. They gave many recitals both public and private, often having musical evenings in their home in Reading. A road accident in 1983 affected Rado's health, and made it impossible for Luise to walk more than a few steps. This sadly diminished their lives. Luise survived Rado by only a few months.
>
> He was the kindest and gentlest of men [98].

We mention that Peter Rado, 1943–2009, was also a mathematician. Peter inherited his parents' talent for music, he was a tubaist and a choral singer.

Erdős and Rado met for the first time in the fall 1934 when Erdős arrived in Cambridge from Budapest. This was the beginning of a lifelong friendship and collaboration. In 1987, in the article entitled *My joint work with Richard Rado*, Erdős recalled:

> I first became aware of Richard Rado's existence in 1933 when his important paper *Studien zur Kombinatorik* appeared. I thought a great deal about the many fascinating and deep unsolved problems stated in this paper, but I never succeeded to obtain any significant results here (. . .) Our joint work extends to more than 50 years; we wrote 18 joint papers (. . .) Our most important work is undoubtedly in set theory and, in particular, the creation of the partition calculus. The term partition calculus is, of course, due to Rado. Without him, I often would have been content in stating only special cases [32].

4.3.2 Rado's Work: Two Examples

Erdős and Rado were both early users and promoters of Ramsey's theorem. For example, in 1950 they started one of their joint articles in the following way:

Apart from its intrinsic interest [Ramsey's theorem] possesses applications in widely different branches of mathematics. Thus in [92] the theorem is used to deal with a special case of the "Entscheidungsproblem" in formal logic. In [28] the theorem serves to establish the existence of convex polygons having any number of vertices when these vertices are to be selected from an arbitrary system of sufficiently many points in a plane. In [91] it is a principal tool in finding all extensions of the distributive law $(a+b)(c+d) = ac+ad+bc+bd$ to the case where the factors on the left-hand side are replaced by convergent infinite series. Finally, Ramsey's theorem at once leads to [Schur's theorem][33].

In what follows we offer two examples of Rado's contributions to the development of Ramsey theory. The first one is of the same spirit as the origin of Schur's theorem: a combinatorial result is used to prove a statement in a seemingly unrelated mathematical context. In the second example we provide a brief introduction to Rado's *arrow notation*.

4.3.2.1 An Application of Ramsey's Theorem

In [91], Rado characterized so-called *distributive sequences*:

Definition 4.3.1 (Distributive Sequence). *We say that a sequence* $(\kappa_\nu, \lambda_\nu)$, $\nu = 1, 2, \ldots$, *of pairs of positive integers is distributive if the convergence of* $\Sigma_{\kappa=1}^{\infty} X_\kappa$ *and* $\Sigma_{\lambda=1}^{\infty} Y_\lambda$ *implies the convergence of the series* $\Sigma_{i=1}^{\infty} X_{\kappa_i} Y_{\lambda_i}$ *and the validity of the equation*

$$\left(\Sigma_{\kappa=1}^{\infty} X_\kappa\right) \cdot \left(\Sigma_{\lambda=1}^{\infty} Y_\lambda\right) = \Sigma_{i=1}^{\infty} X_{\kappa_i} Y_{\lambda_i}.$$

In order to determine all distributive sequences, Rado used Ramsey's theorem to prove the following statement:

Theorem 4.3.1 (Lemma 2 [91]). *For any* $l \in \mathbb{N}$ *there is* $N = N(l) \in \mathbb{N}$ *such that, for any choice of non-empty sets* $M_1, M_2, \ldots, M_N$, *it is always possible to select* l *sets* $M_{a_\lambda} = M'_\lambda$, $(\lambda \leq l)$, *where* $a_1 < a_2 < \ldots < a_l \leq N$, *and to find a number* $p \leq l+1$ *such that, however we choose* $r \leq l$ *numbers* $e_1 < e_2 < \ldots < e_r \leq l$, *we have*

$$M'_{e_1} \cap M'_{e_2} \cap \ldots \cap M'_{e_r} \begin{cases} \neq \emptyset & \text{if } r < p \\ = \emptyset & \text{if } p \leq r \leq l. \end{cases}$$

Proof. Let $l \in \mathbb{N}$. Let $N = N(l)$ be a positive integer with the property that for any l-colouring of $[1, N]^{(l)}$, the set of all l-subsets of $[1, N]$, there is a subset S of $[1, N]$, $|S| = 2l - 1$, such that $S^{(l)}$ is monochromatic. Recall that, by Ramsey's theorem, we can take $N(l) = R_l(l; 2l - 1) = R(\underbrace{l; 2l - 1, \ldots, 2l - 1}_{l})$.

Let $M_1, M_2, \ldots, M_N$ be non-empty sets. We define an l-colouring c of $[1, N]^{(l)}$ in the following way: Let $T = \{x_1, x_2, \ldots, x_l\} \subset [1, N]$, with $x_i < x_{i+1}$, for all $i \in [1, l-1]$. By definition, $c(T) = j \in [1, l]$, if j is the largest number such that $M_{x_1} \cap M_{x_2} \cap \ldots \cap M_{x_j} \neq \emptyset$.

By our choice of N, there is a set $S = \{a_1, a_2, \ldots, a_{2l-1}\} \subset [1, N]$, with $a_i < a_{i+1}$ for all $i \in [1, 2l-2]$, such that $S^{(l)}$ is c-monochromatic. Let $c(T) = j \in [1, l]$, for all $T \in S^{(l)}$, and let $p = j + 1$. Let $M_{a_i} = M'_i$, for $i \in [1, l]$.

Let $r \in [1, l]$ and let $1 \leq e_1 < e_2 < \cdots < e_r \leq l$. If $r = l$ then $\{a_{e_1}, a_{e_2}, \ldots, a_{e_r}\} \in S^{(l)}$. If $r < l$, then $\{a_{e_1}, a_{e_2}, \ldots, a_{e_r}, a_{l+1}, \ldots a_{2l-r}\} \in S^{(l)}$. By the definition of the colouring c, if $r \leq j < p$ then $M'_{e_1} \cap M'_{e_2} \cap \ldots \cap M'_{e_r} \neq \emptyset$, and if $j < p \leq r \leq l$ then $M'_{e_1} \cap M'_{e_2} \cap \ldots \cap M'_{e_r} = \emptyset$. □

4.3.2.2 Rado's Arrow Notation

In the early 1950s, Erdős and Rado introduced what they called the *decomposition relation* $a \to (b_0, b_1)^2$ between cardinals a, b_0, and b_1 [35]. Here, $a \to (b_0, b_1)^2$ holds if, for any set S such that $|S| = a$ and for any colouring $c : S^{(2)} \to \{0, 1\}$, there are $i \in \{0, 1\}$ and a set $S_i \subseteq S$, $|S_i| = b_i$, such that $c\left(S_i^{(2)}\right) = \{i\}$, i.e. for any $T \in S_i^{(2)}$, $c(T) = i$.

For example, the fact that $R(3, 4) = 9$ means that $a \to (3, 4)^2$ holds for any integer $a \geq 9$ and that $8 \to (3, 4)^2$ does not hold.

In a follow-up paper [36], Erdős and Rado extended this notation (and introduced the term *partition relation*) to $a \to (b_0, b_1, \ldots)_k^r$, with a, b_i, r, k being cardinals such that, for each b_i, $i < k$. Here, $a \to (b_0, b_1, \ldots)_k^r$ holds if for any set S such that $|S| = a$ and for any k-colouring c of $S^{(r)}$ there is $i < k$ and a set $S_i \subseteq S$, $|S_i| = b_i$, such that $c\left(S_i^{(r)}\right) = \{i\}$.

If $b_0 = b_1 = \ldots = b$, the convention is to write $a \to (b)_k^r$. If k is a finite cardinal, we write $a \to (b_0, b_1, \ldots, b_{k-1})^r$.

For example, Ramsey's theorem (Theorem 1.3.1) may be stated as:

For every choice of positive integers r, n, μ, there is an integer m_0 such that, if $m \geq m_0$ then $m \to (n)_\mu^r$.

Similarly, it is common to state Theorem 2.4.2, Ramsey's theorem for $\mathbb{N}^{(2)}$ – Two colours, as: $\aleph_0 \to (\aleph_0)_2^2$. Here, $\aleph_0$ represents the smallest infinite cardinal.

András Hajnal and Jean A. Larson on the beginning of the partition calculus:

In the early fifties, Erdős and Richard Rado [35, 36], initiated a systematic investigation of quantitative generalizations of [the Erdős–Rado theorem for pairs]. They called it the partition calculus. There are cases in mathematical history when a well-chosen notation can enormously enhance the development of a branch of mathematics and a case in point is the ordinary partition symbol invented by Rado. It became clear that a careful analysis of the problems according to the size and nature of the parameters leads to an inexhaustible array of problems, each seemingly simple and natural.

A great many new results were proved by the then young researchers. However, unlike many other classical problems, these problems have resisted full solution. The introduction of new methods and the discovery of new ideas usually has given only incremental progress, and objectively, we are as far as ever from complete answers [53].

4.4   RADO'S THEOREM

"This is what I want in heaven... words to become notes and conversations to be symphonies."

Tina Turner, an American-born Swiss singer
and actress

We recall that, by Schur's theorem, any finite colouring of the positive integers contains a monochromatic solution of the equation $x_1 + x_2 - x_3 = 0$. Similarly, by van der Waerden's theorem any finite colouring of the positive integers contains a monochromatic solution of the equation $x_1 + x_2 - 2x_3 = 0$.

Question 4.4.1. Does every two-colouring of the natural numbers contain a monochromatic solution of the equation $x_1 - 2x_2 = 0$?

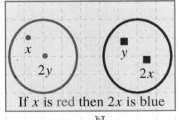

If x is red then $2x$ is blue

Therefore it is possible that an equation does not yield a monochromatic solution for all finite colourings of $\mathbb{N}$. Hence the following definition:

Definition 4.4.1 (Partition Regular Equation). *We say that a homogeneous linear equation $c_1 x_1 + c_2 x_2 + \cdots + c_n x_n = 0$, $c_1, c_2, \ldots, c_n \in \mathbb{Z}$, is partition regular over $\mathbb{N}$ if it has a monochromatic solution whenever $\mathbb{N}$ is finitely coloured.*

As we have seen, the equation $x_1 + x_2 - x_3 = 0$ is partition regular and the equation $x_1 - 2x_2 = 0$ is not.

Example 4.4.1. Check if the equation $x_1 - 2x_2 + 3x_3 = 0$ is partition regular.

Solution. We define a colouring $c : \mathbb{N} \to [1,6]$ in the following way:

$$\text{If } n = 7^k \cdot (7 \cdot l + i), \ i \in [1,6], k,l \geq 0, \text{ then } c(n) = i.$$

In other words, $c(n) = i$ means either that n gives i as the remainder when divided by 7 or n is a product of a power of 7 and a number that gives i as the remainder when divided by 7. See the figure below.

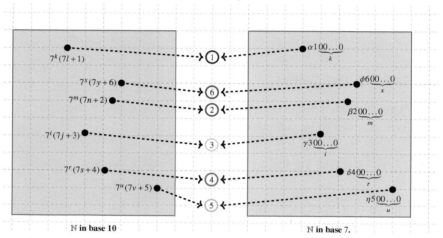

N in base 10   N in base 7.

Observe that, for example, $c(5) = 5$, $c(14) = 2$, $c(25) = 4$, and $c(49) = 1$.

Suppose that there is a c-monochromatic solution of the given equation. Hence suppose that there are $x_1 = 7^k (7l+i)$, $x_2 = 7^s (7t+i)$, and $x_3 = 7^p (7q+i)$, with $i \in [1,6]$ and $k,l,s,t,p,q \geq 0$, such that $x_1 - 2x_2 + 3x_3 = 0$.

It follows that $7^k (7l+i) - 2 \cdot 7^s (7t+i) + 3 \cdot 7^p (7q+i) = 0$, which is the same as, $0 = i \cdot (7^k - 2 \cdot 7^s + 3 \cdot 7^p) + 7 \left(7^k \cdot l - 2 \cdot 7^s \cdot t + 3 \cdot 7^p \cdot q \right)$.

Observe that there is one "extra" factor of 7 in the second term on the right-hand side of the expression above. Hence, if we divide the expression by 7^r, where $r = \min\{k,s,p\}$, then we obtain a contradiction that a number that is not divisible by 7 equals to a number that is divisible by 7.

Since this contradiction is a consequence of our assumption that the colouring c contains a monochromatic solution of the given equation, we conclude that the equation $x_1 - 2x_2 + 3x_3 = 0$ is not partition regular. □

Question 4.4.2. Under which conditions is a homogeneous linear equation $c_1 x_1 + c_2 x_2 + \cdots + c_n x_n = 0$, $c_1, c_2, \ldots, c_n \in \mathbb{Z}$, partition regular over $\mathbb{N}$?

To answer Question 4.4.2 we will use the following tool:

Definition 4.4.2 (*p*-primer).
Let p be a prime number. The
p-primer is a $(p-1)$-colouring
of natural numbers obtained in
the way depicted in the figure
to the right.

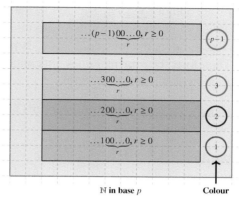

N in base *p*   **Colour**

Observe that the six-colouring used in Example 4.4.1 was the seven-primer.

Proposition 4.4.1. *If integers $c_1, c_2, \ldots, c_n$ are such that, for any subset $J \subseteq [1, n]$, $\sum_{i \in J} c_i \neq 0$ then the equation $c_1 x_1 + c_2 x_2 + \cdots + c_n x_n = 0$ is* NOT *partition regular over* $\mathbb{N}$.

Proof. Let p be a prime such that $p > \sum_{i=1}^{n} |c_i|$ and let $\chi : \mathbb{N} \rightarrow [1, p-1]$ be the p-primer.

Suppose that $X = \{x_1, x_2, \ldots, x_n\}$ is a χ-monochromatic solution of the given equation. Writing $x_1, x_2, \ldots, x_n$ in base p, for some $i \in [1, p-1]$ and $r_1, r_2, \ldots, r_n \geq 0$, we have

$$x_1 = \ldots i \underbrace{0 \cdots 0}_{r_1}, \ x_2 = \ldots i \underbrace{0 \cdots 0}_{r_2}, \ \ldots, \ x_n = \ldots i \underbrace{0 \cdots 0}_{r_n}.$$

Let $r = \min\{r_1, r_2, \ldots, r_n\}$ and $J = \{j \in [1, n] : r_j = r\}$, as shown in the next figure.

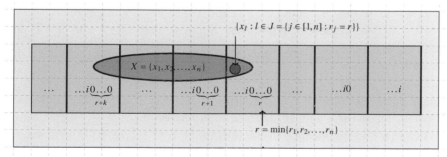

N in base *p* **and the** *p***-primer colour** *i*

Now, if we divide

$$0 = \sum_{j \in J} c_j x_j + \sum_{j \in [1,n] \setminus J} c_j x_j = \sum_{j \in J} c_j \cdot p^r (p k_j + i) + \sum_{j \in [1,n] \setminus J} c_j \cdot p^{r_j} (p k_j + i)$$

by p^r, we obtain $i \sum_{j \in J} c_j + p \cdot A = 0$, for some $A \in \mathbb{Z}$. But this is impossible because, as a product of two non-zero integers with their absolute values less than the prime p, the integer $i \sum_{j \in J} c_j$ is not divisible by p.

Therefore, the p-primer does not contain a monochromatic solution of the equation, which proves the claim. □

Hence, as part of our answer to Question 4.4.2, we have proved the following claim:

Proposition 4.4.2. *Let the equation $c_1 x_1 + c_2 x_2 + \cdots + c_n x_n = 0$ be partition regular over $\mathbb{N}$. Then there is $J \subseteq [1, n]$ such that $\sum_{i \in J} c_i = 0$.*

Observe that Proposition 4.4.2 establishes that the existence of a non-empty set J such that $\sum_{i \in J} c_i = 0$ is a necessary condition for the given equation to be partition regular. In the rest of this section we will show that this condition is also sufficient.

Lemma 4.4.1. *Let $q \in \mathbb{Q}$ and $k \in \mathbb{N}$. Every k-colouring of the natural numbers contains a monochromatic solution of the equation $x + qy = z$.*

Proof. We observe that, for any k-colouring of the natural numbers, the triple $(1, 1, 1)$ is a monochromatic solution of the equation $x + 0 \cdot y = z$.

If $q \neq 0$ and if, for a k-colouring of $\mathbb{N}$, a triple (a, b, c) is a monochromatic solution of the equation $x + qy = z$, then the triple (c, b, a) is a monochromatic solution of the equation $x - qy = z$. Hence we can assume that $q = \frac{r}{s} > 0$, for some $r, s \in \mathbb{N}$.

Our strategy is to use mathematical induction on k, the number of colours, to prove that for any $k \in \mathbb{N}$ there is $n \in \mathbb{N}$ such that any k-colouring of $[1, n]$ contains a monochromatic solution of the equation $x + \frac{r}{s} \cdot y = z$.

If $k = 1$, we take $n = \max\{s, r + 1\}$ and $x = 1$, $y = s$, and $z = r + 1$.

Here is our induction hypothesis: Suppose that $k \geq 1$ and $n \in \mathbb{N}$ are such that any k-colouring of $[1, n]$ contains a monochromatic solution of the equation $x + \frac{r}{s} \cdot y = z$.

Let $W(k + 1; nr + 1)$ be a van der Waerden number, i.e. the least positive integer such that any $(k + 1)$-colouring of $[1, W(k + 1; nr + 1)]$ contains a monochromatic $(nr + 1)$-term arithmetic progression.

Let a $(k + 1)$-colouring c of the interval $[1, W(k + 1; nr + 1)]$ be given.

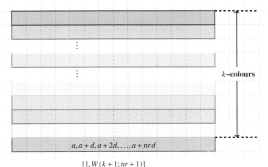

$[1, W(k+1; nr+1)]$

There is a c-monochromatic $(nr+1)$-term arithmetic progression $a, a+d, a+2d, \ldots, a+nrd$.

Next, we consider the set $S = \{ds, 2ds, \ldots, nds\}$ and distinguish two cases:

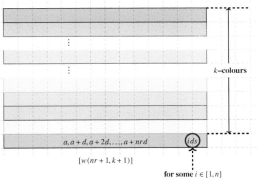

$[w(nr+1, k+1)]$

Case 1: There is $i \in [1, n]$ such that $c(ids) = c(a)$, i.e. for some $i \in [1, n]$, ids is of the same colour class as the monochromatic arithmetic progression.

It follows that, in this case, $(x = a, y = ids, z = a+ird)$ is a monochromatic triple such that $x + \frac{r}{s} y = a + \frac{r}{s} \cdot (isd) = z$.

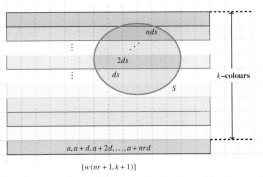

$[w(nr+1, k+1)]$

Case 2: None of the elements of $\{ids : i \in [1, n]\}$ is of the same colour as the monochromatic arithmetic progression. Hence, the set $S = \{ds, 2ds, \ldots, nds\}$ is k-coloured.

We define a k-colouring c' of $[1, n]$ by $c'(i) = c(ids)$, $i \in [1, n]$, and conclude that, by the induction hypothesis, there is a c'-monochromatic triple (a, b, c) such that $a + q \cdot b = c$. It follows that $ads + q \cdot (bds) = cds$ and that, by the definition of the colouring c', the triple (ads, bds, cds) is c-monochromatic.

By the principle of mathematical induction, for any $k \in \mathbb{N}$ there is $n \in \mathbb{N}$ such that any k-colouring of $[1,n]$ contains a monochromatic solution of the equation $x + qy = z$. □

Proposition 4.4.3. *If the set of non-zero integers* $\{c_1, c_2, \ldots, c_n\}$ *is such that* $\sum_{i \in J} c_i = 0$ *for some* $J \subseteq [1,n]$, *then the equation* $c_1 x_1 + c_2 x_2 + \cdots + c_n x_n = 0$ *is partition regular over* $\mathbb{N}$.

Proof. Let χ be a finite colouring of $\mathbb{N}$. Let $i_0 \in J$ and let (a, b, c) be a χ-monochromatic solution of the equation $x + \frac{\sum_{i \in J} c_i}{c_{i_0}} y = z$, that is guaranteed by Lemma 4.4.1.

For $i \in [1, n]$, let

$$\alpha_i = \begin{cases} a & \text{if } i = i_0, \\ b & \text{if } i \notin J, \\ c & \text{if } i \in J \setminus \{i_0\}. \end{cases}$$

Then, since $a c_{i_0} + b \sum_{i \notin J} c_i = c c_{i_0}$, it follows

$$\sum_{i=1}^{n} c_i \alpha_i = a c_{i_0} + b \sum_{i \notin J} c_i + c \sum_{i \in J \setminus \{i_0\}} c_i = c \sum_{i \in J \setminus \{i_0\}} c_i + c c_{i_0} = c \sum_{i \in J} c_i = 0.$$

Therefore, the n-tuple $(\alpha_1, \alpha_2, \ldots, \alpha_n)$ is a χ-monochromatic solution of the given equation.

Since χ was an arbitrarily finite colouring, it follows that the given equation is partition regular over $\mathbb{N}$. □

We summarize Propositions 4.4.2 and 4.4.3 as:

Theorem 4.4.1 (Rado's Theorem [90]). *The equation* $c_1 x_1 + c_2 x_2 + \cdots + c_n x_n = 0$ *is partition regular over* $\mathbb{N}$ *if and only if there is* $J \subseteq [1,n]$ *such that* $\sum_{i \in J} c_i = 0$.

Observe that, by Rado's theorem, to obtain the claim of Schur's theorem it is enough to note that for the first (or the second) and the third coefficient of the equation $x_1 + x_2 - x_3 = 0$ we have $1 + (-1) = 0$. Similarly, by Rado's theorem, to complete Example 4.4.1, i.e. to establish that the equation $x_1 - 2x_2 + 3x_3 = 0$ is not partition regular, it is enough to observe that $1 - 2 = -1 \neq 0$, $1 + 3 = 4 \neq 0$, $1 - 2 + 3 = 2 \neq 0$, and $-2 + 3 = 1 \neq 0$.

4.5 EXERCISES

Exercise 4.1 (*Schur triples and a subset of* $[1,n]$). Let $n > 1$. Let $A \subseteq [1,n]$ be such that $|A| > \frac{n+1}{2}$. Prove that A contains a Schur triple.

Exercise 4.2 (*Schur triples and arithmetic progressions*).

1. Find all red/blue colourings of $\{1,2,3,4,5,6,7\}$ that contain neither a red Schur triple nor a blue three–term arithmetic progression.

2. Show that any red/blue colouring of $\{1,2,3,4,5,6,7,8\}$ contains a red Schur triple or a blue three-term arithmetic progression.

Exercise 4.3 (*Strict Schur triples and arithmetic progressions*). We say that (x,y,z) is a *strict Schur triple* if $x + y = z$ and $x \neq y$. Show that the minimum integer n such that any red/blue colouring of $[1,n]$ must admit either a red strict Schur triple, or a blue three-term arithmetic progression is $n = 10$.

Exercise 4.4 (*Schur's theorem in* $(\mathbb{N},\cdot)$). Prove that any finite colouring of the positive integers admits a monochromatic solution of the equation $xy = z$.

Exercise 4.5 (*Schur's theorem and the equation* $xy + x + y = z$). Prove that any finite colouring of the positive integers admits a monochromatic solution of the equation $xy + x + y = z$.

Exercise 4.6 (*Schur's theorem and the plane* $x + y - z = 0$). Let P be the set of points in the plane $x + y - z = 0$ whose coordinates are positive integers. Let an r-colouring of the positive integers be given. For each $(a,b,c) \in P$, do the following: If a, b, c are of the same colour, then colour the point (a,b,c) by that colour. Otherwise, mark (a,b,c) by an X.

1. Can all of the points of the set P be marked by an X?

2. Can we tell if, under any given finite colouring of the set of positive integers, the set P must contain an infinite number of coloured points?

Exercise 4.7 (*Counting Schur triples*). In this exercise we will explore the following question:

> By Schur's theorem, if $N \in \mathbb{N}$, $N \geq 5$, then, for any two-colouring $c : [1,N] \to \{\circledast, \blacksquare\}$, the set S_c of all c-monochromatic Schur triples is non-empty. What can we say about the number $S(N) = \inf_{c:[1,N] \to \{\circledast, \blacksquare\}} \{|S_c|\}$?

Let $k \in \mathbb{N}$ and let $N = 11k$. Let a two-colouring $\xi : [1, N] \to \{\bullet, \blacksquare\}$ be defined by

$$\xi(n) = \begin{cases} \bullet & \text{if } n \in [1, 4k-1] \cup [10k, 11k = N] \\ \blacksquare & \text{if } n \in [4k, 10k-1]. \end{cases}$$

See below.

In what follows we will count ξ-monochromatic Schur triples $(x, y, x+y)$, $x \le y$.

Let $S_\bullet = \{(x, y, x+y) \in [1, N]^3 : \xi(x) = \xi(y) = \xi(x+y) = \bullet\}$ and $S_\blacksquare = \{(x, y, x+y) \in [1, N]^3 : \xi(x) = \xi(y) = \xi(x+y) = \blacksquare\}$.

1. Show that $|\{(x, y, x+y) \in S_\bullet : x \in [1, k]\}| = \frac{7k^2 - k}{2}$.

2. Show that $|\{(x, y, x+y) \in S_\bullet : x \in [k+1, 2k-1]\}| = k^2 - k$.

3. Determine $|S_\bullet|$.

4. Determine $|S_\blacksquare|$.

5. Determine the number of all ξ-monochromatic Schur triples.

6. Conclude that if $N \in \mathbb{N}$ is divisible by 11 then $|S(N)| < \frac{N^2}{22}$.

Exercise 4.8 (*Generalized Schur number*). The generalized Schur number $s(4, 5)$ is defined as the smallest positive integer n such that any blue/red colouring of the set $[1, n]$ contains a blue solution of the equation $\mathcal{L}(4)$: $x_1 + x_2 + x_3 = x_4$ or a red solution of the equation $\mathcal{L}(5) : x_1 + x_2 + x_3 + x_4 = x_5$.

1. Consider a two-colouring of the interval $[1, 13]$, given by $B = \{1, 2, 12, 13\}$ and $R = [3, 11]$. Check that this colouring does not contain a blue solution of $\mathcal{L}(4)$ or a red solution of $\mathcal{L}(5)$.

2. To show that any blue/red colouring of $[1, 14]$ contains a blue solution of $\mathcal{L}(4)$ or a red solution of $\mathcal{L}(5)$ do the following:

   (a) Suppose that 1 is blue and build a blue/red colouring trying to avoid a blue solution of $\mathcal{L}(4)$ AND a red solution of $\mathcal{L}(5)$.

   (b) Suppose that 1 is red and build a blue/red colouring trying to avoid a blue solution of $\mathcal{L}(4)$ AND a red solution of $\mathcal{L}(5)$.

3. Conclude that $s(4, 5) = 14$.

Exercise 4.9 (*Rainbow Schur triples*). Prove that any equinumerous three-colouring of the set [1,9] contains a rainbow solution of Schur's equation $x + y = z$.

Here, an *equinumerous three-colouring* of [1,9] means that each colour is used exactly three times and *a rainbow solution* means that x, y, and z are all coloured differently.

Exercise 4.10 (*Rado's theorem and partition regular equations*). Are the following equations partition regular over $\mathbb{N}$:

1. $\frac{1}{3}x_1 - \frac{1}{4}x_2 + 2x_3 - \frac{1}{12}x_4 = 0$

2. $\frac{x_1}{6} + \frac{x_2}{4} - \frac{x_3}{12} + \frac{x_4}{6} - \frac{x_5}{3} = 0$

3. $x - y + 5z - 3w = 0$

4. $2x_1 - x_2 + 3x_3 = 6x_4$

Exercise 4.11 (*Rado's theorem and the equation* $x_1 + x_2 - 4x_3 = 0$). Is it possible to find a finite colouring that does not contain monochromatic solutions of the equation $x_1 + x_2 - 4x_3 = 0$? If it is, find such a colouring. If it is not, justify your answer.

Exercise 4.12 (*Folkman's Theorem*). The purpose of this exercise is to establish a proof of Folkman's theorem:

For all $r, k \in \mathbb{N}$ there exists a natural number $M(r,k)$ such that for every r-colouring of $[1, M]$ there exist $a_1, a_2, \ldots, a_k \in [1, M]$, with all a_i distinct, such that the set $F_k(a_1, \ldots, a_k) = \left\{ \sum_{i=0}^{k} \varepsilon_i a_i : \varepsilon_i \in \{0, 1\} \text{ and } \varepsilon_1^2 + \ldots + \varepsilon_k^2 \neq 0 \right\}$ is monochromatic.

In [46], Graham, Rothschild, and Spencer wrote the following: "Although this result was proved independently by several mathematicians, we choose to honour the memory of our friend Jon Folkman [an American mathematician, 1938–1969] by associating his name with this result."[2]

1. What is the set $F_3(1, 2, 5)$?

2. Show that Folkman's theorem is a generalization of Schur's theorem.

3. Prove the following claim by induction on k:

   For all $r, k \in \mathbb{N}$ there exists a natural number $n = n(r,k)$ such that, for any r-colouring χ of $[1, n]$ there exist $a_1, a_2, \ldots, a_k$ with the property that, for any non-empty subset I of the set $[1, k]$, $a(I) = \sum_{i \in I} a_i \in [1, n]$ and $\chi(a(I)) = \chi(a_{\max(I)})$.

   Base Case: Prove the base case, i.e. prove that $n(r, 1)$ exists.

   Induction Step: For the induction step suppose that $k \geq 1$ is such that there exists a natural number $n = n(r, k)$ with the required property.

---

[2]The result is also known as Arnautov–Folkman–Rado–Sanders' Theorem [96].

Let $N = 2W(r; n(r, k) + 1)$, where $W(r; n(r, k) + 1)$ is the van der Waerden number that guarantees the existence of a monochromatic $(n(r, k) + 1)$-term arithmetic progression whenever an interval that contains $W(r; n(r, k) + 1)$ consecutive positive integers is r-coloured.

Fix an r-colouring ξ of the interval $[1, N]$.

(a) Prove that there are $a_{k+1} \in \left[\frac{N}{2} + 1, N\right]$ and $d \in \mathbb{N}$ such that the arithmetic progression $\{a_{k+1} + j \cdot d : 0 \leq j \leq n(r, k)\} \subset \left[\frac{N}{2} + 1, N\right]$ is ξ-monochromatic.

(b) Let d be the number established in (a). Explain why there exist $a_1, a_2, \ldots, a_k \in \{d, 2d, \ldots, n(r, k) \cdot d\}$ such that, for any non-empty subset I of the set $[1, k]$, $a(I) = \sum_{i \in I} a_i \in \{d, 2d, \ldots, n(r, k) \cdot d\}$ and $\xi(a(I)) = \xi(a_{\max(I)})$.

(c) Complete the proof of the induction step.

4. To complete the proof of Folkman's theorem show that one can take $M = M(r, k) = n(r, r(k - 1) + 1)$, where $n(r, r(k - 1) + 1)$ is the number guaranteed by the claim proved in Part 3.

Fix an r-colouring χ of $[1, M]$.

(a) Justify the following claim: There exist $a_1, a_2, \ldots, a_{r(k-1)+1}$ such that, for any non-empty subset I of the set $[1, r(k - 1) + 1]$, $a(I) = \sum_{i \in I} a_i \in [1, M]$ and $\chi(a(I)) = \chi(a_{\max(I)})$.

(b) Define an r-colouring η of the set $[1, r(k - 1) + 1]$ in the following way: For any $j \in [1, r(k - 1) + 1]$, $\eta(j) = \chi(a(I)) = \chi(\sum_{i \in I} a_i)$, where $\emptyset \neq I \subseteq [1, r(k - 1) + 1]$ and $\max(I) = j$.

Is the colouring η well-defined? Why yes, or why not?

(c) Prove that there is an η-monochromatic set $S \subset [1, r(k - 1) + 1]$ such that $|S| = k$.

(d) Finish the proof of Folkman's theorem by proving that the set of all sums of the elements of the set $A = \{a_i : i \in S\}$ is χ-monochromatic.

Exercise 4.13 (*Hilbert's Cube Lemma*). In 1892, David Hilbert, a German mathematician, 1862–1943, one of the most prominent mathematicians of the late nineteenth and early twentieth centuries proved a statement known as Hilbert's cube lemma:

For any r-colouring $\chi : \mathbb{N} \to [1,r]$ and for any $m \in \mathbb{N}$ there exist $a, a_1, \ldots, a_m \in \mathbb{N}$ such that the m-cube $Q_m(a; a_1, \ldots, a_m) = \{a + \sum_{i=0}^{m} \varepsilon_i a_i : \varepsilon_i \in \{0, 1\}\}$ is monochromatic.

Hilbert's cube lemma is considered as the earliest *Ramseyian* theorem.

The purpose of this exercise is to establish the proof of Hilbert's cube lemma in the case $m = 3$. Those readers interested in proving the general case should use the ideas presented below and mathematical induction.

1. Determine all elements of the three-cube $Q(1; 2, 3, 4)$.

2. Recall that χ is the given r-colouring of $\mathbb{N}$. Prove that the interval $[k + 1, k + (r + 1)]$, where $k \in \mathbb{N}$, contains a χ-monochromatic 1-cube.

3. We say that a χ-monochromatic one-cube $Q(a; a_1) \subset [k + 1, k + (r + 1)]$, where $k \in \mathbb{N}$, is of the type (a_1, i) if $\chi(Q(a; a_1)) = \{i\}$. How many different types of χ-monochromatic one-cubes in $[k + 1, k + (r + 1)]$, an interval of length $r + 1$, are possible?

4. Observe that the interval $[1, (r^2 + 1)(r + 1)]$ is the union of $r^2 + 1$ consecutive intervals of length $r + 1$, $[1, (r^2 + 1)(r + 1)] = \cup_{i=0}^{r^2}[i(r + 1) + 1, (i + 1)(r + 1)]$.

   (a) Prove that there are p and q, $0 \le p \le q \le r^2$, such that the intervals $[p(r+1)+1, (p+1)(r+1)]$ and $[q(r+1)+1, (q+1)(r+1)]$ contain χ-monochromatic one-cubes of the same type.

   (b) Prove that the interval $[1, (r^2 + 1)(r + 1)]$ contains a χ-monochromatic two-cube.

5. Observe that, from Part 4, it follows that any interval $[k + 1, k + (r^2 + 1)(r + 1)]$, where $k \in \mathbb{N}$, contains a χ-monochromatic two-cube $Q(a; a_1, a_2)$, with $a_1 \in [1, r]$. We say that a χ-monochromatic two-cube $Q(a; a_1, a_2) \subset [k + 1, k + (r^2 + 1)(r + 1)]$, where $k \in \mathbb{N}$ and $a_1 \in [1, r]$, is of the type (a_1, a_2, i) if $\chi(Q(a; a_1, a_2)) = \{i\}$. Establish that the number of possible types of χ-monochromatic two-cubes in $[k + 1, k + (r^2 + 1)(r + 1)]$ is less than $(1 + r)^5$.

6. Prove that the interval $[1, (r^2 + 1)(r + 1)^6]$ contains a χ-monochromatic three-cube.

CHAPTER 5

The Hales–Jewett Theorem

The Hales–Jewett THEOREM is another of the landmarks in the development of Ramsey theory.

The theorem was inspired by Rado's notion of regularity, van der Waerden's theorem, and a generalization of a well-known children's game. In their article *Regularity and positional games* [54], Hales and Jewett demonstrated that "the concept of regularity is useful in analyzing certain types of games."

It turned out that the publication of [54] was just, to paraphrase William Moser, a Canadian mathematician, 1927–2009, the beginning of the Hales–Jewett theorem's long and happy mathematical life.

5.1 COMBINATORIAL LINES

> "Last year I went fishing with Salvador Dali. He was using a dotted line. He caught every other fish."
>
> Steven Wright, an American comedian,
> actor, and writer

In this section we introduce the notion of a *combinatorial line*. In what follows we will justify the *line* part in the name of these intriguing objects, but also provide the evidence that a combinatorial line is something very different from a Euclidean line.

Definition 5.1.1 (Alphabet). *For $m \in \mathbb{N}$, any set A such that $|A| = m$ is called an* alphabet *on m symbols.*

DOI: 10.1201/9781003286370-5

Example 5.1.1. Let $A = \{a, 1, \triangle\}$. Then A is an alphabet on $|A| = 3$ symbols.

Definition 5.1.2 (Words). *Let A be an alphabet on m symbols. For $n \in \mathbb{N}$, any function $w : [1, n] \to A$ is called a* word *of length n on the alphabet A. If $w(i) = a_i$, $i \in [1, n]$, then we write $w = a_1\, a_2 \cdots a_n$.*

The set of all words of length n on the alphabet A is denoted by A^n. We say that A^n is *the n-dimensional cube on the alphabet A.*

Example 5.1.2. Let $A = \{a, 1, \triangle\}$ be an alphabet on three symbols. Then $w = a\, 1\, a\, 1\, a\, 1$ is a word of length six on the alphabet A. Here $w : [1, 6] \to A$ is defined as $w(1) = w(3) = w(5) = a$ and $w(2) = w(4) = w(6) = 1$.
   Also, $A^2 = \{a\, a, a\, 1, a\, \triangle, 1\, a, 1\, 1, 1\, \triangle, \triangle\, a, \triangle\, 1, \triangle\, \triangle\}$.

Definition 5.1.3 (Roots). *Let A be an alphabet (on m symbols) and let $*$ be a symbol such that $* \notin A$.[1] We consider the alphabet $A_* = A \cup \{*\}$. Any word on the alphabet A_*, i.e. any element of $(A_*)^n = A_*^n$, for some $n \in \mathbb{N}$, that contains the symbol $*$ is called a* root.

Example 5.1.3. Let $A = \{a, 1, \triangle\}$ be an alphabet on three symbols. Then $A_* = A \cup \{*\} = \{a, 1, \triangle, *\}$. By definition, $1\, *\, \triangle \in A_*^3$ and $a\, *\, a\, * \in A_*^4$ are roots and the word $1\, 1\, \triangle \in A_*^3$ is not.

For a root $\tau \in A_*^n$ and a symbol $a \in A$ we define the word $\tau_a \in A^n$ in the following way: For $i \in [1, n]$,

$$\tau_a(i) = \begin{cases} \tau(i) & \text{if} \quad \tau(i) \neq *, \\ a & \text{if} \quad \tau(i) = *. \end{cases}$$

Example 5.1.4. Let $A = \{a, b, c\}$ and let $\tau = *\, b\, c\, b \in A_*^4$ be a root. Then, $\tau_a = a\, b\, c\, b$, $\tau_b = b\, b\, c\, b$, and $\tau_c = c\, b\, c\, b$.

Example 5.1.5. Let $A = [1, 4]$ and let $\tau = *\, 1\, 3\, *\, 4\, * \in A_*^6$ be a root. Then $\tau_2 = 2\, 1\, 3\, 2\, 4\, 2$.

Definition 5.1.4 (Combinatorial Line). *Let A be an alphabet, let $n \in \mathbb{N}$, and let $\tau \in A_*^n$ be a root. A* combinatorial line *in A^n rooted in τ is the set of words $L_\tau = \{\tau_a : a \in A\}$.*

We observe that $L_\tau \subseteq A^n$, for any root $\tau \in A_*^n$.

Example 5.1.6. Let $A = [1, 3] = \{1, 2, 3\}$ and $n = 2$. Find all combinatorial lines in A^2.

---

[1]The symbol $*$ is commonly called a *wildcard* or an *active coordinate*.

Solution. Observe that the set of all roots in A^2_* is given by $\{\tau = * \, 1, \sigma = * \, 2, \theta = * \, 3, \rho = 1 \, *, \chi = 2 \, *, \phi = 3 \, *, \mu = * \, *\}$.

It follows that all combinatorial lines in A^2 are given by: $L_\tau = \{1 \, 1, 2 \, 1, 3 \, 1\}$, $L_\sigma = \{1 \, 2, 2 \, 2, 3 \, 2\}$, $L_\theta = \{1 \, 3, 2 \, 3, 3 \, 3\}$, $L_\rho = \{1 \, 1, 1 \, 2, 1 \, 3\}$, $L_\chi = \{2 \, 1, 2 \, 2, 2 \, 3\}$, $L_\phi = \{3 \, 1, 3 \, 2, 3 \, 3\}$, and $L_\mu = \{1 \, 1, 2 \, 2, 3 \, 3\}$.

Here is another view of all combinatorial lines in A^2.

L_τ	L_σ	L_θ	L_ρ	L_χ	L_ϕ	L_μ
1 1	1 2	1 3	1 1	2 1	3 1	1 1
2 1	2 2	2 3	1 2	2 2	3 2	2 2
3 1	3 2	3 3	1 3	2 3	3 3	3 3

For yet another view of all combinatorial lines in $A^2 = [1,3]^2$, we first observe the one-to-one correspondence between the two-dimensional cube on alphabet $[1,3]$ and the set of points $P = \{(x, y) : x, y \in [1,3]\}$ in the xy-plane.

By this correspondence, each combinatorial line corresponds to a line segment in the xy-plane that contains three points from the set P.

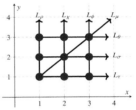

Observe that the south-east diagonal of the square determined by the set P does not correspond to any of the combinatorial lines in $[1,3]^2$. □

Finally, we observe that each of the combinatorial lines in $[1,3]^2$ corresponds to a winning position in the Tic-Tac-Toe game.

Tic-Tac-Toe: ✕ wins! Tic-Tac-Toe: ● **wins!**

Again, observe that there is a winning position, the south-east diagonal, that does not correspond to any of the combinatorial lines in $[1,3]^2$.

Example 5.1.7. Determine combinatorial lines in $[1,3]^3$ rooted in $\tau = * \, 2 \, 3$, $\sigma = * \, * \, 3$, and $\theta = * \, * \, * \in [1,3]^3_*$.

Solution. By definition:

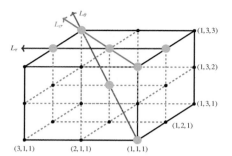

L_τ	L_σ	L_θ
1 2 3	1 1 3	1 1 1
2 2 3	2 2 3	2 2 2
3 2 3	3 3 3	3 3 3

We observe that the points in $\mathbb{R}^3$ that correspond to the elements of the combinatorial line L_τ lie on the Euclidean line ℓ_τ with the parametric equations $x = t, y = 2, z = 3, t \in \mathbb{R}$. Similarly, the combinatorial line L_σ corresponds to a set of points on the line ℓ_σ with the parametric equations $x = t, y = t, z = 3, t \in \mathbb{R}$, and L_θ corresponds to a set of points on the line ℓ_θ with the parametric equations $x = t, y = t, z = t, t \in \mathbb{R}$. □

In general, for $m \in \mathbb{N}$, the combinatorial line determined by a root $\mu = a_1\, a_2\, a_3 \in [1, m]_*^3$ corresponds to a set of points on the Euclidean line ℓ_μ with the parametric equations $x = b_1 + \alpha_1 \cdot t, y = b_2 + \alpha_2 \cdot t, z = b_3 + \alpha_3 \cdot t, t \in \mathbb{R}$, where $b_i = 0$ and $\alpha_i = 1$ if $a_i = *$, and $b_i = a_i$ and $\alpha_i = 0$ if $a_i \in [1, m]$. This set of points is obtained for the values $t \in [1, m]$.

5.2 GENERALIZED TIC-TAC-TOE GAME

"The ambition should always be to play an elegant game."

Edson Arantes do Nascimento Pelé, a Brazilian
footballer

1940–2022

Recall the game of Tic-Tac-Toe: two players take turns claiming the spaces in a 3×3 grid with the goal to claim a row, a column, or a diagonal.

A group of undergraduate students created a web game called *Quad-Tac-Toe* [4]. This is a player vs. AI game on a $4 \times 4 \times 4$ cube $Q(4, 3) = \{(x, y, z) : x, y, z \in [1, 4]\}$. The player and AI are given two different colours and tasked to colour one point in $Q(4, 3)$ at each turn. Who first completes a monochromatic line wins.[2]

---

[2]In 1980, this variant of the game of Tic-Tac-Toe was in detail analyzed by Oren Patashnik, an American computer scientist [86].

Here, *a line* means a set of four collinear points, either horizontally, vertically, or diagonally.

Quad-Tac-Toe

We illustrate the fact that each of the combinatorial lines in $[1,4]^3$ corresponds to a winning position in the Quad-Tac-Toe game by the following example:

For $a, b \in [1,4]$, consider the root $\tau = * \, a \, b \in [1,4]^3_*$ and the Euclidean line ℓ_τ in $\mathbb{R}^3$ given by its parametric equations $x = t, y = a, z = b, t \in \mathbb{R}$. Hence, ℓ_τ is the line that passes through the point $(0, a, b)$ and is parallel to the x-axis. Now, the combinatorial line $L_\tau = \{1 \, a \, b, 2 \, a \, b, 3 \, a \, b, 4 \, a \, b\}$ corresponds to the winning position $\{(1, a, b), (2, a, b), (3, a, b), (4, a, b)\} = \{(t, a, b) : t \in [1,4]\} \subseteq \ell_\tau \cap Q(4, 3)$.

Observe that the line $x = t, y = 5 - t, z = 1, t \in \mathbb{R}$, contains the winning position $\{(1, 4, 1), (2, 3, 1), (3, 2, 1), (4, 1, 1)\} = \{(t, 5 - t, 1) : t \in [1,4]\}$ that does not correspond to any of the combinatorial lines in $[1,4]^3$.

Actually, there are 76 different Euclidean lines containing four points from $Q(4, 3)$, but only 61 of them correspond to combinatorial lines in the three-dimensional cube on alphabet $[1,4]$. Hence, not every winning position in the Quad-Tac-Toe game corresponds to a (monochromatic) combinatorial line. See Exercises 5.4. and 5.5.

But what if one considers a k-player game played on the "board" $Q(m, n) = \{(x_1, x_2, \ldots, x_n) : x_1, x_2, \ldots, x_n \in [1, m]\}$?[3] Similar to the Tic-Tac-Toe game and the Quad-Tac-Toe game, each player is given one of k different colours and tasked to colour one point in $Q(m, n)$ at each turn. The player who first completes a monochromatic line wins. Here "a line" means a set of m collinear points, i.e. a set of m points in $Q(m, n)$ that lie on a line in $\mathbb{R}^n$ given by its parametric equations $x_1 = a_1 + \alpha_1 \cdot t, x_2 = a_2 + \alpha_2 \cdot t, \ldots, x_n = a_n + \alpha_n \cdot t, t \in \mathbb{R}$, for some fixed real numbers $a_i, \alpha_i, i \in [1, n]$.

Graham, Rothschild, and Spencer called the above generalization of Tic-Tac-Toe a "k-person n-dimensional Tic-Tac-Toe m-in a row" game [46].

---

[3]Observe that $(x_1, x_2, \ldots, x_n) \in Q(m, n)$ is a point in $\mathbb{R}^n$ with $x_1, x_2, \ldots, x_m \in [1, m] = \{1, 2, \ldots, m\}$. By $[1, m]^n = \{x_1 \, x_2 \, \cdots \, x_n : x_1, x_2, \ldots, x_n \in [1, m]\}$ we denote the n-cube on the alphabet $[1, m]$.

Proposition 5.2.1. *Any combinatorial line in* $[1,m]^n$ *corresponds to a winning position in a k-person n-dimensional Tic-Tac-Toe m-in a row game.*

Proof. For given $m, n \in \mathbb{N}$, we consider a root $\tau = a_1 a_2 \ldots a_n \in [1,m]_*^n$. Let the line ℓ_τ in $\mathbb{R}^n$ be given by its parametric equations

$$x_1 = b_1 + \alpha_1 \cdot t, x_2 = b_2 + \alpha_2 \cdot t, \ldots, x_n = b_n + \alpha_n \cdot t, t \in \mathbb{R},$$

where $b_i = 0$ and $\alpha_i = 1$ if $a_i = *$, and $b_i = a_i$ and $\alpha_i = 0$ if $a_i \in [1,m]$.

Recall that for $i \in [1,m]$, the word $\tau_i = x_1^{(i)} x_2^{(i)} \cdots x_n^{(i)} \in [1,m]^n$ is such that $x_j^{(i)} = i$ if $a_j = *$ and $x_j^{(i)} = a_j$ if $a_j \in [1,m]$.

It follows that, by taking $t = i$ in the parametric equation for ℓ_τ, the point $(x_1^{(i)}, x_2^{(i)}, \ldots, x_n^{(i)}) \in \ell_\tau \cap Q(m,n)$.

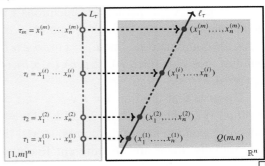

This establishes an injection between the combinatorial line $L_\tau = \{\tau_i : i \in [1,m]\}$ and the set $\ell_\tau \cap Q(m,n)$. Since $|L_\tau| = m$, the combinatorial line L_τ corresponds to a winning position. □

Suppose that a k-person n-dimensional Tic-Tac-Toe m-in a row game ended up in a tie. Hence, the set $Q(m,n)$ was partitioned into k mutually disjunct parts, $Q(m,n) = P_1 \cup \ldots \cup P_k$, $P_i \cap P_j = \emptyset$ if $i \neq j$, in such a way that none of the parts contained a set of m collinear points.

Let us define a k-colouring C of the cube $[1,m]^n$ in the following way: for $a_1 \cdots a_n \in [1,m]^n$, $C(a_1 \cdots a_n) = i$ if and only if $(a_1, \ldots, a_k) \in P_i$. By Proposition 5.2.1 and our assumption that the game ended up in a tie, it follows that the colouring C of the n-dimensional cube on alphabet $[1,m]$ does not contain a monochromatic combinatorial line.

In the spirit of Ramsey theory, this observation leads us to the following question:

Question 5.2.1. Let A be an alphabet on m symbols and let A^n be the n-dimensional cube on alphabet A, i.e. let $A^n = \{a_1 a_2 \cdots a_n : a_i \in A, i \in [1,n]\}$. If A^n is k-coloured, under which conditions can we be sure that A^n contains a monochromatic combinatorial line?

More precisely, let $k, m \in \mathbb{N}$ and let A be an alphabet on m symbols.

Does there exist an $n \in \mathbb{N}$ such that whenever A^n is k-coloured there exists a monochromatic line?

$\{♥, ♣, ♠, ♦\}^n$ is k–coloured. . .   . . . and a red **combinatorial line.**

A nice answer to Question 5.2.1 would be that, for any $k, m \in \mathbb{N}$ and for a big enough natural number n, the k-person n-dimensional Tic-Tac-Toe m-in a row game cannot end up in a tie.

This is exactly what Hales and Jewett discussed in their paper *Regularity and position games*, published in 1963 [54]. Their famous Hales–Jewett theorem establishes that, if the dimension is sufficiently large, a generalized Tic-Tac-Toe game never ends up in a tie.

Alfred W. Hales and Robert I. Jewett:

Alfred Washington Hales and Robert Israel Jewett had long and distinguished academic careers, Hales at the University of California Los Angeles and Jewett at the University of Western Washington.

When they submitted the *Regularity and position games* paper in 1961, Hales was 23 years old and Jewett was 24. Both were doctoral students at the time. Hales was working under the supervision of Robert P. Dilworth at the California Institute of Technology (Caltech) and Jewett was supervised by Karl Stromberg at the University of Oregon.

The pair knew each other from their time as undergraduate students at Caltech.

In 1971, Hales and Jewett, together with Ron Graham, Klaus Leeb, and Bruce Rothschild, were the first recipients of the George Pólya Prize.

5.3   THE HALES–JEWETT THEOREM

"The Hales–Jewett theorem strips van der Waerden's theorem of its unessential elements and revels the heart of Ramsey theory."

Ron Graham, Bruce Rothschild, and Joel Spencer

In the proof of the Hales–Jewett theorem presented in this section, similar to the proof of van der Waerden's theorem, we will again follow Leader and use a notion of the so-called *focused* and *colour-focused* combinatorial lines [74].

Theorem 5.3.1 (The Hales–Jewett Theorem [54]). *Let $k, m \in \mathbb{N}$ and let A be an alphabet on m symbols. There exists an $n \in \mathbb{N}$ such that whenever A^n is k-coloured there exists a monochromatic combinatorial line.*

Definition 5.3.1 (Hales–Jewett Number). *The smallest n guaranteed by Theorem 5.3.1 is denoted by $H(k; m)$ and called a Hales–Jewett number.*

Let $k, m \in \mathbb{N}$. For the rest of this chapter, as an alphabet on m symbols, we will take the set $A = [1, m]$.

Definition 5.3.2 (Focused Combinatorial Lines). *Let $r \in \mathbb{N}$ and let $\tau^{(1)}, \tau^{(2)}, \ldots, \tau^{(r)} \in [1, m]_*^n$ be r roots. We say that the corresponding combinatorial lines are focused at $f \in [1, m]^n$ if $\tau_m^{(1)} = \tau_m^{(2)} = \cdots = \tau_m^{(r)} = f$.*

Example 5.3.1. Consider $\tau^{(1)}, \tau^{(2)}, \tau^{(3)} \in [1, 4]_*^4$ given by $\tau^{(1)} = * * 3 *$, $\tau^{(2)} = * 4\ 3 *$, and $\tau^{(3)} = * * 4\ 3\ 4$.

Then, from $\tau_4^{(1)} = 4\ 4\ 3\ 4$, $\tau_4^{(2)} = 4\ 4\ 3\ 4$, and $\tau_4^{(3)} = 4\ 4\ 3\ 4$, it follows that the corresponding combinatorial lines are focused at $f = 4\ 4\ 3\ 4$:

$L_{\tau^{(1)}}$	$L_{\tau^{(2)}}$	$L_{\tau^{(3)}}$
1 1 3 1	1 4 3 1	1 4 3 4
2 2 3 2	2 4 3 2	2 4 3 4
3 3 3 3	3 4 3 3	3 4 3 4
4 4 3 4	4 4 3 4	4 4 3 4

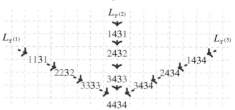

Let c be a k-colouring of $[1, m]^n$ and let $\tau^{(1)}, \tau^{(2)}, \ldots, \tau^{(r)} \in [1, m]_*^n$ be r roots.

Definition 5.3.3 (Colour-focused Combinatorial Lines). *We say that r combinatorial lines $L_{\tau^{(1)}}, L_{\tau^{(2)}}, \ldots, L_{\tau^{(r)}}$ are colour-focused if:*

1. *For each $i \in [1, r]$, $c(\tau_1^{(i)}) = c(\tau_2^{(i)}) = \cdots = c(\tau_{m-1}^{(i)})$.*

2. *For each $i, j \in [1, r]$, if $i \neq j$ then $c(\tau_1^{(i)}) \neq c(\tau_1^{(j)})$.*

3. *Combinatorial lines $L_{\tau^{(1)}}, L_{\tau^{(2)}}, \ldots, L_{\tau^{(r)}}$ are focused at some $f \in [1, m]^n$.*

Note distinct colours and the common focus $\tau_m^{(1)} = \tau_m^{(2)} = \cdots = \tau_m^{(r)}$ in the figure below.

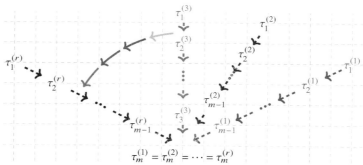

Our next goal is to prove Theorem 5.3.1. Recall that $H(k;m)$ denotes the least integer n such that whenever $[1,m]^n$ is k-coloured there exists a monochromatic combinatorial line. Observe that, at this point, we have not proved that $H(k;m)$ exists for all $k,m \in \mathbb{N}$. Similar to our proof of van der Waerden's theorem, to prove the Hales–Jewett theorem we will use double induction.

Proof of Theorem 5.3.1. The main strategy of the proof is to use induction on m, the size of the alphabet $A = [1,m]$, to establish that $H(k;m)$ exists for all $k, m \in \mathbb{N}$.

If $m = 1$ then $A = \{1\}$. Hence, for any $n \in \mathbb{N}$, $A^n = \{1\,1\,\cdots\,1\}$. This means that there is only one combinatorial line in A^n, the single-element cube itself. This implies that any k-colouring of A^n will contain a monochromatic combinatorial line. Hence, $H(k;1) = 1$. This completes the proof of the base case of induction.

Let $m > 1$. We assume that $H(k;m-1)$ exists for all k.

We fix $k \in \mathbb{N}$.

To complete the induction step we need the following claim:

Claim 5.3.1. *For all $1 \le r \le k$, there exists n such that whenever $[1,m]^n$ is k coloured, there exists* either *a monochromatic combinatorial line or r colour-focused combinatorial lines.*

Proof of Claim 5.3.1. We use induction on r, the number of colour-focused combinatorial lines. Recall that we are assuming that m is such that $H(k;m-1)$ exists for all k.

To prove that the claim is true if $r = 1$, we take $n = H(k;m-1)$.

Let c be a k-colouring of $[1,m]^n$. The colouring c induces a k-colouring of $[1,m-1]^n$. Our choice of n guarantees the existence of a monochromatic line in $[1,m-1]^n$. Hence, there is a c-monochromatic combinatorial line *or* one colour-focused combinatorial line in the cube $[1,m]^n$.

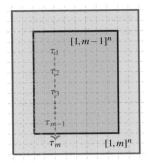

Where are you?

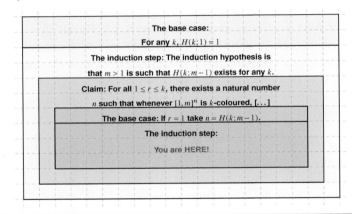

For the induction step, let $r \in [1, k-1]$ and let, now, $n = n(r)$ be such that whenever $[1,m]^n$ is k coloured, there exists *either* a monochromatic combinatorial line *or* r colour-focused combinatorial lines. Let $n' = HJ(k^{m^n}; m-1)$ and let $N = n + n'$.[4]

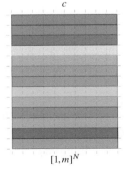

Suppose that a k-colouring c of $[1,m]^N = [1,m]^{n+n'}$ *does not contain a c-monochromatic combinatorial line.*

---

[4]The reason why we need $HJ(k^{m^n}; m-1)$ will become clear shortly.

$c_b(a) = c(ab)$

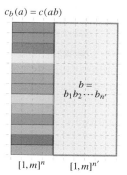

$[1,m]^n$ $[1,m]^{n'}$

$\chi(b) = c_b$

Next, for $b = b_1 b_2 \cdots b_{n'} \in [1,m]^{n'}$ we define c_b, a k-colouring of $[1,m]^n$, so that, for $a \in [1,m]^n$, $c_b(a) = c(ab)$. In other words, via the colouring c, to any $b = b_1 b_2 \cdots b_{n'} \in [1,m]^{n'}$ we associate a k-colouring of the n-cube $[1,m]^n$.

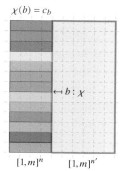

$[1,m]^n$ $[1,m]^{n'}$

Recall that there are k^{m^n} distinct k-colourings of the cube $[1,m]^n$. Hence, the mapping $\chi : b \mapsto c_b$ is a k^{m^n}-colouring of the n'-cube $[1,m]^{n'}$.

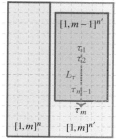

$[1,m]^N = [1,m]^{n+n'} = [1,m]^n \times [1,m]^{n'}$

By our choice of n', there is a χ-monochromatic line in $[1,m-1]^{n'}$, i.e. there is a root $\tau \in [1,m-1]_*^{n'} \subset [1,m]_*^{n'}$ such that the combinatorial line $L_\tau \subseteq [1,m-1]^{n'}$ is χ-monochromatic. We observe that the line $L_\tau \subseteq [1,m]^{n'}$ is χ-colour-focused with the focus τ_m.

c_τ

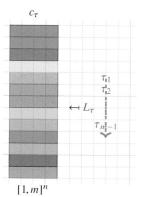

$[1,m]^n$

The phrase "L_τ is χ-monochromatic" means that each element of the combinatorial line $L_\tau \subset [1,m-1]^{n'}$ induces (via the k-colouring c) the same k-colouring of the n-cube $[1,m]^n$. We call this colouring c_τ. It follows that, for any $a \in [1,m]^n$, $c_\tau(a) = c(a\tau_1) = c(a\tau_2) = \cdots = c(a\tau_{m-1})$.

Recall that we have assumed that c, a k-colouring of the $(n + n')$-cube $[1, m]^{n+n'}$, does not contain a monochromatic combinatorial line. This assumption implies that c_τ, a k-colouring of the n-cube $[1, m]^n$, does not contain a monochromatic combinatorial line. Otherwise, if $\sigma \in [1, m]_*^n$ is a root such that the combinatorial line L_σ is c_τ-monochromatic then, for the root $\sigma' = \sigma\tau_1 \in [1, m]_*^{n+n'}$, the combinatorial line $L_{\sigma'}$ would be c-monochromatic, which contradicts our assumption.

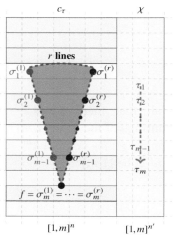

By our choice of n, there are r c_τ-colour-focused combinatorial lines $L_{\sigma^{(1)}}, \ldots,$ $L_{\sigma^{(r)}}$ in the n-cube $[1, m]^n$ with the focus f. None of the combinatorial lines $L_{\sigma^{(1)}},$ $\ldots, L_{\sigma^{(r)}}$ is monochromatic. Recall that the combinatorial line $L_\tau \subset [1, m]^{n'}$ is χ-colour-focused with the focus τ_m.

Next we define $r + 1$ roots in $[1, m]_*^N$ as follows: $\tau^{(1)} = \sigma^{(1)}\tau$, $\tau^{(2)} = \sigma^{(2)}\tau, \ldots, \tau^{(r)} = \sigma^{(r)}\tau$, $\tau^{(r+1)} = f\tau$.

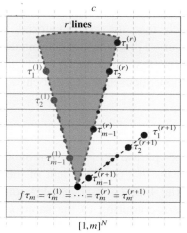

Recall that, by our definition of the colouring c_τ, it follows that $c(\tau_j^{(i)}) = c(\sigma_j^{(i)}\tau_j) = c_\tau(\sigma_j^{(i)})$, for any $i \in [1, r]$ and any $j \in [1, m-1]$. Similarly, $c(\tau_j^{(r+1)}) = c(f\tau_j) = c_\tau(f)$, for any $j \in [1, m-1]$.

Hence, there are $r + 1$ c-coloured-focused lines $L_{\tau^{(1)}}, \ldots, L_{\tau^{(r+1)}}$ in the N-cube $[1, m]^N$ with the focus $f\tau_m$.

We have established that $N = n + n' \in \mathbb{N}$ is such that any c-colouring of the cube $[1, m]^N$ contains or a c-monochromatic combinatorial line or $r + 1$ colour-focused combinatorial lines, which completes the induction step and the proof of Claim 5.3.1. $\qquad\square$

Where are you?

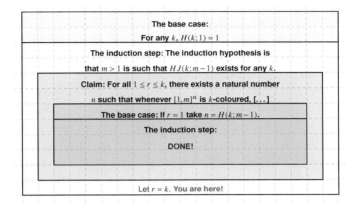

The base case:

For any k, $H(k; 1) = 1$

The induction step: The induction hypothesis is
that $m > 1$ is such that $HJ(k; m-1)$ exists for any k.

Claim: For all $1 \le r \le k$, there exists a natural number
n such that whenever $[1, m]^n$ is k-coloured, [. . .]

The base case: If $r = 1$ take $n = H(k; m-1)$.

The induction step:

DONE!

Let $r = k$. You are here!

By Claim 5.3.1, there is $n = n(k) \in \mathbb{N}$ such that any k-colouring c of $[1, m]^n$ contains a c-monochromatic combinatorial line or k colour-focused combinatorial lines.

If we suppose that a k-colouring c of $[1, m]^n$ does not contain a monochromatic combinatorial line, then there must be k colour-focused combinatorial lines. But, since each of those k combinatorial lines uses a different colour, one of the colours must match the colour of the focus.

A monochromatic combinatorial line appears! This contradicts our assumption that the colouring c does not contain a monochromatic combinatorial line.

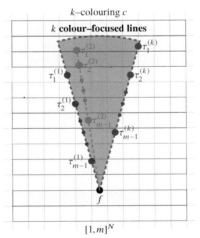

Therefore, for any $m > 1$ such that $H(k; m-1)$ exists for all $k \in \mathbb{N}$, for any k we can find a number $n = n(k)$ such that every k-colouring of the n-cube $[1, m]^n$ contains a monochromatic combinatorial line. This completes the induction step and the proof of the theorem. □

Example 5.3.2. Use the Hales–Jewett theorem to prove van der Waerden's theorem: If $k, l \in \mathbb{N}$ then any l-colouring of $\mathbb{N}$ contains a k-term monochromatic arithmetic progression.

Solution. Let $k, l \in \mathbb{N}$ be given. Let $c : \mathbb{N} \to [1, l]$ be an l-colouring of the set of natural numbers. Let $N = HJ(l; k)$.

We define an l-colouring of the N-cube $[1, k]^N$ as follows: if $x_1 x_2 \cdots x_N \in [1, k]^N$ then $c'(x_1 x_2 \cdots x_N) = c(x_1 + x_2 + \ldots + x_N)$.

By the Hales–Jewett theorem there is a c'-monochromatic combinatorial line L_τ rooted in a root $\tau = a_1 a_2 \cdots a_N \in [1, k]_*^N$. We observe that there is at least one $i \in [1, N]$ such that $a_i = *$.

Let $S = \{i \in [1, N] : a_i \in [1, k]\}$. We recall that $L_\tau = \{\tau_1, \tau_2, \ldots, \tau_k\}$ where, for $j \in [1, k]$, $\tau_j = a_1^{(j)} a_2^{(j)} \cdots a_N^{(j)}$, with $a_i^{(j)} = a_i$ if $i \in S$ and $a_i^{(j)} = j$ if $i \notin S$. Let $a = \sum_{i \in S} a_i$ and let $d = |[1, N] \backslash S|$.

Note that, for $j \in [1, k]$, $\sum_{i=1}^{N} a_i^{(j)} = \sum_{i \in S} a_i^{(j)} + \sum_{i \in [1, N] \backslash S} a_i^{(j)} = a + dj$.

On the other hand, $c'(\tau_1) = c'(\tau_2) = \cdots = c'(\tau_k)$ which together with, for each $j \in [1, k]$, $c'(\tau_j) = c\left(\sum_{i=1}^{N} a_i^{(j)}\right) = c(a + jd)$, implies $c(a + d) = c(a + 2d) = \cdots = c(a + kd)$.

Hence, the k-term arithmetic progression $a + d, a + 2d, \ldots, a + kd$ is c-monochromatic. □

The Hales–Jewett Theorem and the Polymath Project

Similar to Ramsey's theorem, van der Waerden's theorem, Schur's theorem, and Rado's theorem, the Hales–Jewett theorem was one of those significant mathematical results that became a source of inspiration for generations of mathematicians.

Here we briefly reflect on one of the developments, closely related to the Hales–Jewett theorem, that has enriched the whole 21st century's mathematical landscape.

In 1983, Graham offered $1,000 for a proof of what he called, a "density version for the Hales–Jewett theorem" (DHJ):

For all finite A and $\varepsilon > 0$ there exists $N(A, \varepsilon)$ such that if $N \geq N(A, \varepsilon)$ and $R \subseteq A^N$ satisfies $|R| \geq \varepsilon |A^N|$ then R must contain a combinatorial line [47].

In 1991, Furstenberg and Yitzhak Katznelson, an Israeli mathematician, proved Graham's conjecture from 1983 by using the ergodic theory techniques [41].

In the early 2009, in a series of blog posts, Gowers invited members of the mathematical community to jointly search for an elementary proof of DHJ. Gowers's initial blog titled *Is massively collaborative mathematics possible?* [43] marks the beginning of the Polymath Project, a still ongoing collaboration of mathematicians across the world on a variety of important mathematical problems.

For all of those who study, teach, and do mathematics, here are two of the "ground rules" of the Polymath Project that Gowers established in 2009:

"3. When you do research, you are more likely to succeed if you try out lots of stupid ideas. Similarly, stupid comments are welcome here. (In the sense in which I am using "stupid," it means something completely different from "unintelligent." It just means not fully thought through.)

5. Don't *actually* use the word "stupid," except perhaps of yourself [43]."

The first Polymath Project was a success. In 2012, the first elementary proof of DHJ, together with a quantitative bound on how large n needs to be, was published [89]. The collaborators fittingly named the author "D.H.J. Polymath".

5.4 EXERCISES

Exercise 5.1 (*Combinatorial lines*). Let $A = \{a, b, c, d\}$. List all combinatorial lines in A^2.

Exercise 5.2 (*Combinatorial lines*). In the figure to the right you see four Euclidean lines passing through points in a $4 \times 4 \times 4$ cube. If a line corresponds to a combinatorial line in $[1,4]^3$, list all of the elements of the combinatorial line and determine the root associated with it. If a line does not correspond to a combinatorial line, briefly explain why.

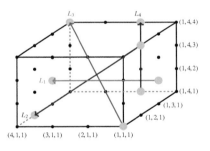

Exercise 5.3 (*Focused combinatorial lines*). Let $A = \{1, 2, 3, 4, 5\}$. Give an example of five focused combinatorial lines in A^5.

Exercise 5.4 (*How many Euclidean lines are there?*). In the May 1947 issue of The American Mathematical Monthly, A. L. Rubinoff from the University of Toronto proposed the following problem: Suppose that a noughts and crosses (i.e. a Tic-Tac-Toe game) are played on an n-dimensional cube of side k. Show that there are precisely $\frac{(k+2)^n - k^n}{2}$ rows, columns, diagonals . . . on which a win may be scored [99].

Exercise 5.5 (*How many combinatorial lines are there?*). Let $m, n \in \mathbb{N}$ and let $|A| = m$, i.e. let A be an alphabet on m symbols. Prove that the number of combinatorial lines in A^n equals to $(m + 1)^n - m^n$.

Exercise 5.6 (*No monochromatic lines – no contradiction*). Let $A = \{a, b, c, d\}$. Can you find a two-colouring of A^2 that does not contain a monochromatic combinatorial line? If yes, does this contradict the claim of the Hales–Jewett theorem?

Exercise 5.7 (*Is it true that $HJ(2;3) = 2$?*). Let $A = \{1, 2, 3\}$ be an alphabet. Check if the two-colouring depicted in the figure to the right yields a monochromatic combinatorial line in A^3. Based on your observation, what can you tell about the Hales–Jewett number $H(2;3)$?

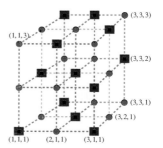

Exercise 5.8 (*Hales–Jewett numbers*). Prove that, for $r \geq 2$, $HJ(r;2) = r$.

Exercise 5.9 (*Monochromatic powers of 2*). Use the Hales–Jewett theorem to prove that for any two-colouring of natural numbers there is an ℓ-term arithmetic progression $a_1, a_2, \ldots, a_\ell$, not necessarily monochromatic, such that the set $A = \{2^{a_1}, 2^{a_2}, \ldots, 2^{a_\ell}\}$ is monochromatic.

Exercise 5.10 (*Gallai's theorem for semigroups*). Tibor Gallai, a Hungarian mathematician, 1912–1992, was a lifelong friend and collaborator of Paul Erdős. In this exercise we use the Hales–Jewett theorem to prove Gallai's theorem for semigroups.

Let $(A, \bullet)$ be a semigroup and let $\ell \in \mathbb{N}$ be given. For any r-colouring of the set A there are $a, b \in A$ such that the set $\{a, a \bullet b, a \bullet b^2, \ldots, a \bullet b^{\ell-1}\}$ is monochromatic.

Note: A semigroup $(A, \bullet)$ is an algebraic structure consisting of a set A together with an associative binary operation $\bullet$. Commonly, $x \bullet y$ denotes the result of applying the semigroup operation to the ordered pair (x, y). Associativity is expressed as $(x \bullet y) \bullet z = x \bullet (y \bullet z)$, for all $x, y, z \in A$.

For $k \in \mathbb{N}$ and $x \in A$ we write

$$x^k = \underbrace{x \bullet (x \bullet (x \bullet (\ldots (x \bullet x) \ldots)))}_{k} = \underbrace{x \bullet x \bullet \ldots x \bullet x}_{k}.$$

For example, both $(\mathbb{N}, +)$ and $(\mathbb{N}, \cdot)$ are semigroups.

Exercise 5.11 (*Gallai–Witt theorem*). Ernst Witt, 1911–1991, was a German mathematician. In this exercise we will use the Hales–Jewett theorem to prove the Gallai–Witt theorem:

Let $(V, +, \cdot)$ be a real vector space and let $A = \{a_1, a_2, \ldots, a_m\}$ be a finite subset of V. Prove that, for any r-colouring of the set V, there exist a vector $u \in V$ and a real number λ such that the set $u + \lambda \cdot A = \{u + \lambda \cdot a_1, u + \lambda \cdot a_2, \ldots, u + \lambda \cdot a_m\}$ is monochromatic.

CHAPTER 6

Happy End Problem

$\text{Erd\H{o}s}$ AND SZEKERES rediscovered Ramsey's theorem while looking for a solution to the Happy End Problem. It turned out that the problem of finding convex n-gons, attributed to *Miss Klein* in [28], had multiple happy endings: Esther Klein and George Szekeres got married and lived happily together for 68 years; Paul Erdős found a mathematical area that would interest him throughout his life; a beautiful piece of mathematics was created; a gigantic step towards the establishment of Ramsey theory was made; the field of Euclidean Ramsey theory was initiated; and, for many decades, the Happy End Problem itself has been studied in higher dimensions and generalized in the various mathematical settings.

Still, at the time of writing this text, the conjecture, stated by Erdős and Szekeres in 1935, that any $2^{n-2} + 1$ points in general position must contain the vertices of a convex n-gon has not been completely resolved.

In this chapter, we will provide details about several classical results related to the Happy End Problem and briefly summarize the Problem's 90-year history.

6.1   THE HAPPY END PROBLEM: TRIANGLES, QUADRILATERALS, AND PENTAGONS

"Where there is love there is life."

Mahātmā Gandhi, Indian leader

1869–1948

In this section we will discuss a few special cases of what Erdős and Szekeres called *a combinatorial problem in geometry* [28], the problem of

DOI: 10.1201/9781003286370-6

finding conditions under which, for a given natural number n, a set of points in the plane must contain a set of vertices of a convex n-gon. We will briefly mention the case $n = 3$, remind the reader what happens if $n = 4$, and discuss in detail the case $n = 5$.

As a warm-up we start with the following question:

Question 6.1.1. Is it true that any finite set with more than two points must contain three points that are the vertices of a triangle?

The negative answer to Question 6.1.1 motivates the following definition:

Definition 6.1.1. *We say that A is a set of points in general position if there is no line that contains three points from A.*

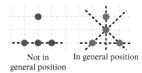

Not in general position   In general position

Clearly, any set of at least three points in general position contains three points that are the vertices of a triangle. But what if one asks for the existence of a convex n-gon for $n > 3$?

Problem 6.1.1 (Esther Klein's Problem on Convex n-gons [28]). *Can we find for a given n a number $N(n)$ such that from any set containing at least N points in general position it is possible to select n points forming a convex polygon?*

We have established that $N(3) = 3$.

We remind the reader that in 1934, the year when Erdős and Szekeres submitted their manuscript that introduced Problem 6.1.1 to the world, Erdős was only 21 years old and Szekeres, at 23, was only a bit older. Still, as the true trailblazers, they asked two questions that would become part of the very core of Ramsey theory:

(1) Does the number N corresponding to n exist? (2) If so, how is the least $N(n)$ determined as a function of n? (...) We obtain a certain preliminary answer to the second question. But the answer is not final for we generally get in this way a number N which is too large [28].

Definition 6.1.2 (Erdős–Szekeres Number). *For a given $n \geq 3$, we say that $ES(n)$ is an Erdős–Szekeres number if it is the least among all natural numbers N with the property that any N points in general position contain vertices of a convex n-gon.*

Hence, $ES(3) = 3$. To proceed with finding $ES(4)$ and $ES(5)$ we need the following definition:

Definition 6.1.3 (Convex Hull). *The convex hull of a set of points S in the plane is the smallest convex set $\overline{S}$ that contains S.*

If the set S is bounded then it is common to visualize its convex hull as the part of the plane bounded by the final position of a rubber band that is initially stretched to contain the set S and then released.

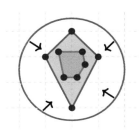

For example, consider the set S with nine points, like one in the figure to the right. Stretch a rubber band around the set S and release it. Observe that rubber band's final position will be the boundary of $\overline{S}$, a convex quadrilateral, a 4-gon. Also, observe that the convex hull of the remaining five points is a convex pentagon, a 5-gon. We will say that the set S is of type $(4,5)$.

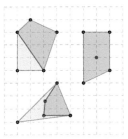

In Section 1.2, we told a story in which Esther Klein demonstrated to her friends that $ES(4) = 5$. See the figure to the right for three possible cases: types $(5,0)$, $(4,1)$, and $(3,2)$. Also observe that the convex hull of a set of four points that is of type $(3,1)$ is a triangle.

For the rest of this section we are concerned with the case $n = 5$.

Lemma 6.1.1. $ES(5) > 8$.

Proof. We consider the set of all vertices of two nested rectangular quadrilaterals.

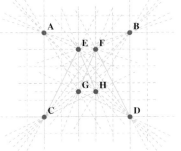

We choose a set of eight points in general position of type $(4,4)$, $S = \{A, B, C, D, E, F, G, H\}$, so that the line segments $\overline{AD}$ and $\overline{BC}$ intersect both $\overline{EG}$ and $\overline{EH}$ but do not intersect either $\overline{EF}$ or $\overline{GH}$. We claim that S does not contain vertices of a convex pentagon.

Let X be any five-element subset of S and let $S' = \{A, B, C, D\}$ and $S'' = \{E, F, G, H\}$.

By the pigeonhole principle there are four possible cases, depending on whether $|X \cap S'| = 1, 2, 3$, or 4.

If $|X \cap S'| = 4$ then the convex hull of X is the rectangle with the vertices A, B, C, and D. Hence X is of type $(4, 1)$.

Let $|X \cap S'| = 3$, say $\{A, B, C\} \subset X$. Then $X \cap S'' = \{M, N\} \subset \{E, F, G, H\}$. If M (or N) belongs to the interior of $\triangle ABC$ then the convex hull of X is a triangle or a quadrilateral. In the remaining case, $\{M, N\} = \{G, H\}$, the convex hull is the quadrilateral with the vertices A, B, H, and C.

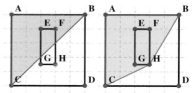

Let $|X \cap S''| = 3$, say $\{E, F, G\} \subset X$. Then $X \cap S' = \{M, N\} \subset \{A, B, C, D\}$.

Suppose that the line segment $\overline{MN}$ is parallel to the line segment $\overline{EF}$ or to the line segment $\overline{EG}$. Observe that in this case X contains vertices of a trapezoid.

If the fifth element of X belongs to the interior of this trapezoid, i.e. if $\{M, N\} = \{B, D\}$ or $\{M, N\} = \{C, D\}$, then the convex hull of X is a quadrilateral.

Otherwise, $\{M, N\} = \{A, B\}$ or $\{M, N\} = \{A, C\}$ and the convex hull of X is a triangle or a quadrilateral, i.e. X is of type $(3,2)$ or $(4, 1)$.

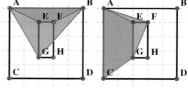

If the line segment $\overline{MN}$ is parallel neither to the line segment $\overline{EF}$ nor to the line segment $\overline{EG}$, i.e. if $\{M, N\} = \{A, D\}$ or $\{M, N\} = \{B, C\}$, then the convex hull of X is a quadrilateral, i.e. X is of type $(4, 1)$.

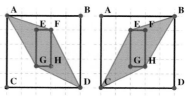

Hence, if $|X \cap S''| = 3$ then X is of type $(3, 2)$ or $(4, 1)$.

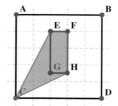

In the last case, $|X \cap S''| = 4$, the convex hull of X is a quadrilateral, i.e. X is of type $(4, 1)$.

Therefore, we constructed a set of eight points in general position that does not contain the vertices of a convex pentagon. In other words, $ES(5) > 8$. □

Lemma 6.1.2. $ES(5) \leq 9$.

Proof. Let $S = \{A, B, C, D, E, F, G, H, I\}$ be a set of nine points in the plane in general position. Let $\overline{S}$ be the convex hull of S.

If $\overline{S}$ has *five or more* vertices, then any five of those vertices determine a convex pentagon. Observe that this confirms that if S is of type $(x, *)$, with $x \in \{5, 6, 7, 8, 9\}$, then S contains vertices of a convex pentagon.

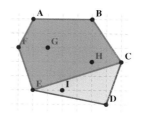

Let $\overline{S}$, the convex hull of S, has three or four vertices. Then the set $T = S \backslash \overline{S}$ contains six or five (remaining) points of S and they are all inside $\overline{S}$. Let $\overline{T}$ be the convex hull of T.

If $|\overline{T}| = 5$ or $|\overline{T}| = 6$ then any five of those vertices determine a convex pentagon. Observe that this confirms that if S is of type $(x, y, *)$, with $x \in \{3, 4\}$, $y \in \{5, 6\}$, and $x + y \le 9$, then S contains vertices of a convex pentagon.

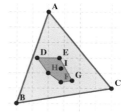

There are four types remaining: $(3, 3, 3)$, $(3, 4, 2)$, $(4, 3, 2)$, and $(4, 4, 1)$.

We start our analysis of type $(3, 3, 3)$ by considering the eight-point configuration of type $(3, 3, 2)$.

We observe that, in the inner configuration $(3, 2)$, the line that contains the line segment must intersect two sides of the triangle. We notice the vertex where those two sides of the triangle meet and draw four rays starting at the end points of the line segment as in the figure to the right.

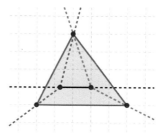

Notice three open regions in the plane in the figure to the right. We observe the following: (a) None of three regions intersects the interior of the triangle; (b) Region 1 and Region 2 intersect each other (part of the plane 'above' the top vertex); (c) Region 3 does not intersect either Region 1 or Region 2.

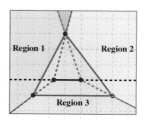

Note that the remaining three points (the vertices of the outside triangle) in the configuration $(3, 3, 2)$ cannot be on the boundary of any of Regions 1–3.

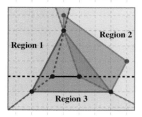

If one of the outside points belongs to Region 3, then a convex pentagon appears.

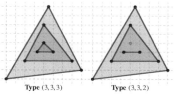

If none of the outside points belongs to Region 3, then, by the pigeonhole principle, two of the outside points belong to the same region, say to Region 2. A convex pentagon appears again.

Hence a configuration of type $(3, 3, 2)$ contains vertices of a convex pentagon.

Since the configuration of type $(3, 3, 3)$ contains the configuration $(3, 3, 2)$, the configuration $(3, 3, 3)$ contains vertices of a convex pentagon.

Next we consider the eight-point configuration of type $(4, 3, 1)$.

We consider the inner configuration $(3, 1)$, a triangle and a single point inside of it. Through each vertex we draw a ray with its initial point at the inside point and note three regions, as shown in the figure.

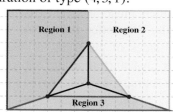

By the pigeonhole principle, at least two of the remaining four points must belong to the same region, say Region 2. A convex pentagon appears.

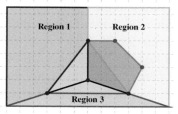

Hence a configuration of type $(4, 3, 1)$ contains vertices of a convex pentagon.

Note that configurations $(4,4,1)$ and $(4,3,2)$ contain the configuration $(4,3,1)$. It follows that configurations of types $(4,4,1)$ and $(4,3,2)$ contain vertices of a convex pentagon.

Finally, we analyze the configuration of type $(3,4,2)$. We consider the inside configuration $(4,2)$, a quadrilateral and a line segment.

Suppose that the line that contains the line segment intersects the adjacent sides of the quadrilateral. A convex pentagon appears!

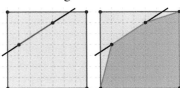

Now, suppose that the line that contains the line segment intersects the opposite sides of the quadrilateral. We draw four rays from the end points of the line segment through the vertices of the quadrilateral. These rays divide the part of the plane outside of the quadrilateral into four disjunct regions, R_1, R_2, R_3, and R_4. See the figure to the right.

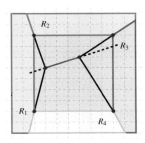

If one of the outside three points belongs to R_2 or R_4 then a convex pentagon appears.

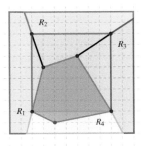

If none of the three outside point belongs to R_2 or R_4 then, by the pigeonhole principle, at least two points must belong to R_1 or R_3. A convex pentagon appears again.

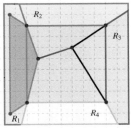

It follows that a configuration of type $(3,4,2)$ contains vertices of a convex pentagon.

Therefore we have proved that $ES(5) \leq 9$. □

From Lemma 6.1.1 and Lemma 6.1.2 it follows that

Theorem 6.1.1. $ES(5) = 9$.

Back to the Városliget Park in the early 1930s:

In 2021, Endre Makai Jr., a Hungarian mathematician and the oldest son of Endre Makai, 1915–1987, wrote to a group of Canadian undergraduate students:

"My father and Paul Turán together made [in 1934] probably the first proof of the problem of Erdős and Szekeres in the case of nine points. Namely, any nine points in the plane, no three lying on a line, contain as a subset all five vertices of a convex pentagon. They did not publish this result, but later there appeared several different proofs for this special case [78]."

The proof presented in this section follows the proof by William E. Bonnice, an American mathematician, from [10].

6.2   THE HAPPY END PROBLEM: GENERAL CASE

"I have always considered my work a joint effort. I was fortunate to have worked on great ideas and with very intelligent people. I may have developed a few equations no one had thought of before but that was nothing unusual-everybody did that."

Mary G. Ross, A Cherokee engineer, and the first female
engineer in the history of Lockheed

1908–2008

In this section we will prove that $ES(n)$ exists for any $n \geq 3$. We will follow Erdős–Szekeres' paper [28] and provide two proofs of this fact.

Recall that we have established that $ES(3) = 3 = 2^1 + 1 = 2^{3-2} + 1$, $ES(4) = 5 = 2^2 + 1 = 2^{4-2} + 1$, and $ES(5) = 9 = 2^3 + 1 = 2^{5-2} + 1$. Moreover, Szekeres and Lindsay Peters, an Australian scientist and software developer, proved that $ES(6) = 17 = 2^4 + 1 = 2^{6-2} + 1$ [116].

Hence this thrilling conjecture stated by Erdős and Szekeres in 1935:

Conjecture 6.2.1. *For any $n \geq 3$, $ES(n) = 2^{n-2} + 1$.*

Not long before his death in 1996, Erdős wrote that he would pay $500 for a proof of this conjecture.

Still... How do we prove that $ES(n)$ exists for all $n \geq 3$?

6.2.1 Proof Via Ramsey's Theorem

Recall that, for any natural number n, Ramsey's theorem establishes the existence of the Ramsey number $R(4; n, 5)$, a number with the following property: if S_m is a set with m elements, $m \geq R(4; n, 5)$, and if the set of all four-element subset of S_m is two-coloured then there exists an n-element subset $\Delta_n \subseteq S_m$ such that all its four-element subsets are coloured by the first colour, or there exists a five-element subset $\Delta_5 \subseteq S_m$ such that all its four-element subsets are coloured by the second colour.

In what follows we will use the following fact:

Lemma 6.2.1. *If $n \geq 4$ then n points in the plane form a convex polygon if and only if every four of them form a convex quadrilateral.*

Proof. We use mathematical induction to prove the lemma. The base case $n = 4$ is trivial.

Suppose that the claim is true for some fixed $n \geq 4$, i.e. suppose that $n \geq 4$ is such that any n points in the plane form a convex polygon if and only if every four of them form a convex quadrilateral.

Let S be a set of $n + 1$ points in the plane. If those points form a convex $(n + 1)$-gon then any four points from S, as vertices of a convex polygon, form a convex quadrilateral.

Suppose that any four points from S form a convex quadrilateral.[1] We note that by the induction hypothesis this means that any n points from S form a convex n-gon.

Let $\overline{S}$ be the convex hull of S. Then $|\overline{S}| = n + 1$ or $|\overline{S}| = n$.

Suppose that $|\overline{S}| = n$. Then there is a point $A \in S$ that is inside the n-gon formed by points from $\overline{S}$. Let $X \in \overline{S}$ and consider the line through the points A and X. This line intersects the line segment with endpoints Y and Z, for some $Y, Z \in S$.

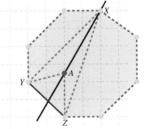

Observe that, since the point A is inside of $\triangle XYZ$, the quadrilateral determined by the points A, X, Y, and Z is not convex. This contradicts our assumption that any four points from S form a convex quadrilateral.

Hence $|\overline{S}| = n + 1$ and S is a convex $(n + 1)$-gon, which completes the induction step. $\qquad\qquad\square$

---

[1]Observe that this implies that S is a set of points in general position.

Theorem 6.2.1 (The Happy End Problem Via Ramsey's Theorem). $ES(n)$ *exists for any* $n \geq 3$.

Proof. Let $n \geq 4$, let $m \geq R(4; n, 5)$, and let S_m be a set of m points in the plane in general position. Let $S_m^{(4)} = \{\{A, B, C, D\} : A, B, C, D \in S_m\}$, i.e. let $S_m^{(4)}$ be the set of all four-element subsets of S_m.

We define a two-colouring $c : S_m^{(4)} \to \{■, •\}$ in the following way:

For $T \in S_m^{(4)}$, let $T_\square$ be the quadrilateral with T as its set of vertices. Then:

$$c(T) = \begin{cases} ■ & \text{if } T_\square \text{ is convex,} \\ • & \text{if } T_\square \text{ is not convex,} \end{cases}$$

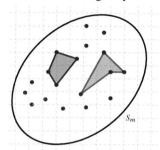

By Ramsey's theorem (and the fact that m is greater than or equal to the Ramsey number $R(4; n, 5)$), there is an n-element set $\Delta_n \subset S_m$ such that all of its four-element subsets are coloured ■ or a five-element set $\Delta_5 \subset S_m$ such that all of its four-element subsets are coloured •.

Since $ES(4) = 5$, any set of five points in the plane in general position contains a convex quadrilateral. This implies that it is impossible to find a five-element set $\Delta_5 \subset S_m$ such that all of its four-element subsets are coloured •.

Hence there must be an n-element set $\Delta_n \subset S_m$ such that all of its four-element subsets are coloured ■. But then, by Lemma 6.2.1, Δ_n is the set of vertices of a convex n-gon. $\qquad\square$

Observe that we have established that, for any $n \geq 4$, $ES(n) \leq R(4; n, 5)$.

6.2.2 Proof Via Theorem on Cups and Caps

Assume that a rectangular coordinate system in the plane is given. Let $T = \{(x_i, y_i) : i \in [1, n]\}$ be a set of points in general position in the first quadrant such that $x_i < x_{i+1}$, for all $i \in [1, n-1]$.

Definition 6.2.1. We say that T is an n-cup if the sequence of slopes corresponding to line segments in the polygonal line determined by the set T is increasing. The set T is an n-cap if the sequence of slopes is decreasing.

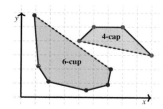

Observe that if we find a cup or a cap in some set of points, then we will also find the set of vertices of a convex polygon.

It is possible that a given set of points in a particular coordinate system is neither cup or cap. See the figure to the right and observe the polygonal line contains a line segment that is parallel to the vertical axis.

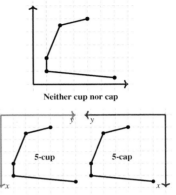

Neither cup nor cap

We also observe that, depending on the choice of the coordinate system, the same set of points can be a cup, a cap, or neither cup nor cap.

Let $S = \{A_1, \ldots, A_n\}$ be a set of points in general position.

Definition 6.2.2 (Right Coordinate System). *We say that a rectangular coordinate system is* right *for S if S is contained in the first quadrant and if there are no two points in S that belong to a vertical line.*

Observe that, since the set S is finite, we can always choose a rectangular coordinate system such that the vertical axis is not parallel to any of the line segments $\overline{A_i A_j}$, $i \neq j$, $i, j \in [1, n]$, and that the horizontal axis is chosen so that the set S is contained in the first quadrant.

In other words, for any finite set S of points in general position there is a rectangular coordinate system that is right for S.

Suppose that S is the set of vertices of a convex n-gon. Observe that if a rectangular coordinate system is right for S then the point with the lowest x-coordinate and the point with the highest x-coordinate are the end points of a cup or a cup contained in S.

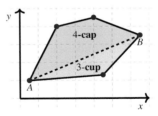

Definition 6.2.3 (Number $f(k, l)$). *For $k, l \geq 3$ we define $f(k, l)$ to be the least positive integer such that any set S of points in general position such that $|S| \geq f(k, l)$ contains, in a coordinate system that is right for S, either a k-cup or an l-cap.*

Observe that for any set S, a k-cup and an l-cap in one system that is right for S may become a k-cap and an l-cup in another coordinate system that is right for S. Hence, $f(k, l) = f(l, k)$.

Example 6.2.1. Prove that for any $k \geq 3$, $f(k,3) = f(3,k) = k$.

Solution. Let $k \geq 3$ and let S be a set of k points in the plane in general position in a coordinate system that is right for S. Let $S = \{A_1, A_2, \ldots, A_k\}$ where, for $i < j$, the x-coordinate of A_i is less than the x-coordinate of A_j. Let s_i be the slope of the line segment $\overline{A_i A_{i+1}}$, for $i \in [1, k-1]$.

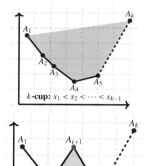

If $s_1 < s_2 < \cdots < s_{k-1}$, then S determines a k-cup.

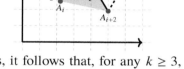

If, for some $i \in [1, k-2]$, $s_i > s_{i+1}$, then the set S contains a 3-cap $\{A_i, A_{i+1}, A_{i+2}\}$.

Since S was an arbitrarily set of k points, it follows that, for any $k \geq 3$, $f(k,3) = f(3,k) = k$. □

Theorem 6.2.2 (Theorem on Cups and Caps). *For any $k, l \geq 3$ the number $f(k,l)$ exists and, for $k, l \geq 4$, $f(k,l) \leq f(k-1,l) + f(k,l-1) - 1$.*

Proof. We prove the theorem by induction on $m = k + l \geq 6$.

Observe that Example 6.2.1 implies that $f(3,3) = 3$ and $f(4,3) = f(3,4) = 4$. In particular this means that if $k, l \geq 3$ are such that $k + l \leq 7$ then $f(k,l)$ exists. This establishes the base case of induction.

Let $m \geq 7$ be such that whenever $u + v = m$ and $u, v \geq 3$ then $f(u,v)$ exists.

We choose $k, l \geq 3$ such that $k + l = m + 1$. This implies that $(k-1) + l = k + (l-1) = m$ and, hence $f(k-1,l)$ and $f(k,l-1)$ exist.

Let $n = f(k-1,l) + f(k,l-1) - 1$.

We fix a set S of n points in general position and a right for S system of coordinates. We have to prove that S contains either a k-cup or an l-cap.

Let L be the set of all points that are the left ends of $(k-1)$-cups in S. In the figure to the right, for $k-1 = 4$, the top left point (■) belongs to L. Observe that if we remove ■, then we are left only with $(k-2) = 3$-cups.

Let us assume first that the set $S\backslash L$ has at least $f(k-1,l)$ points. Then $S\backslash L$ contains either a $(k-1)$-cup or an l-cap. But taking the set L out of S, including the case $L = \emptyset$, leaves no $(k-1)$-cups in $S\backslash L$. Hence, $S\backslash L$ must contain an l-cap.

The remaining case is when $S\backslash L$ has *at most* $f(k-1,l) - 1$ points. Since S has $n = f(k-1,l) + f(k,l-1) - 1$ points, in this case, there must be at least $f(k,l-1)$ points in L. Hence, by Definition 6.2.3, L contains either a k-cup or an $(l-1)$-cap. We observe that, since $L \subseteq S$, the existence of a k-cup in L implies the existence of a k-cup in S.

Suppose that there is no a k-cup in L. Since $|L| \geq f(k,l-1)$ then there must be an $(l-1)$-cap in L. Let us consider the point Y, the right end of that cap.

Let X be the point that is immediately to the left of Y in the $(l-1)$-cap in L. Since $Y \in L$, the point Y is the left end of some $(k-1)$-cup in S. Let Z be the point that is immediately to the right of Y in the $(k-1)$-cup in S.

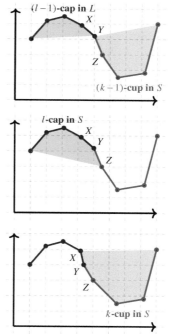

If the slope of $\overline{XY}$ is greater than the slope of $\overline{YZ}$, then adding the point Z to the $(l-1)$-cap in L makes an l-cap in S.

If the slope of $\overline{XY}$ is smaller than the slope of $\overline{YZ}$, then adding the point X to the $(k-1)$-cup in S makes a k-cup in S.

Therefore, any set S such that $|S| = n = f(k-1,l) + f(k,l-1) - 1$ contains either a k-cup or an l-cup which implies that the number $f(k,l)$ exists and completes the induction step.

By the principle of mathematical induction, $f(k,l)$ exists for any $k,l \geq 3$. □

The solution to the Happy End Problem is an immediate consequence of Theorem 6.2.2: any set S of $f(n,n)$ points in general position contains an n-cap or an n-cup, i.e. contains the set of vertices of a convex n-gon.

Esther Szekeres–Klein, 1910–2005, and George Szekeres, 1911–2005:

Esther Klein and George Szekeres got married in 1937 in Budapest, Hungary, which was the reason that Erdős referred to the problem of finding convex *n*-gons as *The Happy End Problem*.

Described by their contemporaries as "a wonderful and unpretentious couple," Esther and George lived full and rich lives. Their love for each other, their family, friends, mathematics, and music were some of the passions that Esther and George shared throughout seven decades of their life together.

In 1939, to escape the war in Europe, Esther and George fled from their beloved Budapest to Shanghai, China. Many years later, George explained: "In Budapest we couldn't think of [having a child]; the incredibly uncertain atmosphere there in the late '30s was not one where a child would want to have to be born. But we were optimists! In '39 I was 28 years old, a very optimistic age: 'Let's have a child,' and that was it, no matter whether we could afford to bring up a child! So, at the first opportunity my son Peter was born, in '40, a year after we got to Shanghai [59]."

As evidence that mathematics may play a role of an anchor amid turbulent times, we mention that Esther Szekeres of Shanghai, China, proposed the following problem in the October 1946 issue of The American Mathematical Monthly: "Let there be given five points in the plane. Prove that we can select four of them which determine a convex polygon [114]."[2]

In 1948, the Szekeres family moved to Adelaida, Australia. It is remembered that "George and Esther quickly fell in love with the Australian bush. Esther's time was taken up with raising their young family, though she tutored in mathematics at the university. George flourished as a professional mathematician [18]."

George made significant contributions to several mathematical areas, ranging from combinatorics to number theory to complex analysis. And still, he is perhaps the best known for his contribution to understanding black holes in cosmology.

Esther and George died within one hour of each other on August 28, 2005.

6.3   ERDŐS–SZEKERES' UPPER AND LOWER BOUNDS

"And then we try to tighten the bounds - to be more precise in terms of how big is big when we say big."

Amanda Montejano, a Mexican mathematician

---

[2]There is another problem proposed by Erdős in the same issue of The Monthly. One may wonder if Erdős in fact submitted the problem in Esther's name.

We start this section by introducing a class of sets of points in general position that we will use to demonstrate both lower and upper bounds of $ES(n)$ obtained by Erdős and Szekeres.

Let a rectangular coordinate system in the plane be given. For $k, l \geq 3$, we say that a set S of points in general position has the $(k, l)^+$-property, if it satisfies the following conditions:

1. The given coordinate system is right for S.

2. $|S| = \binom{k+l-4}{k-2}$.

3. S has no k-cups or l-caps;

4. All lines that contain two points from S have positive slopes.

Example 6.3.1. The set $U = \{(1,1), (2,2)\}$ is with the $(3,3)^+$-property, the set $V = \{(3,1), (4,2), (5,4)\}$ is with the $(4,3)^+$-property, and the set $W = \{(6,1), (7,3), (8,4)\}$ is with the $(3,4)^+$-property.

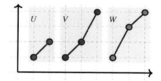

Proposition 6.3.1. *For any $k, l \geq 3$ there is a set S with the $(k, l)^+$-property.*

Proof. We prove the lemma by induction on $m = k + l \geq 6$. The base case of induction is established in Example 6.3.1.

Let $m \geq 7$ be such that, whenever $u + v = m$ and $u, v \geq 3$, there is a set S with the $(u, v)^+$-property.

We choose $k, l \geq 3$ such that $k + l = m + 1$. This implies that $(k-1) + l = k + (l-1) = m$ and, by the induction hypothesis, there is a set A with the $(k-1, l)^+$-property and a set B with the $(k, l-1)^+$-property .

Since $|A| = \binom{k+l-5}{k-3}$ and $|B| = \binom{k+l-5}{k-2}$, i.e. since both sets A and B are finite, the set of slopes of lines passing through two points from A or two points from B is finite. Hence, if necessary, we can translate the set B to obtain a set B' with the following properties:

1. Every point in B' has a greater first coordinate than any point in A.

2. The slope of any line through a point in A and a point in B' is greater than the slope of any line through two points in the same set.

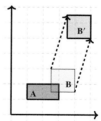

We consider the set $S = A \cup B'$. Since A and B' are disjoint sets, it follows that

$$|S| = |A| + |B'| = \binom{k+l-5}{k-3} + \binom{k+l-5}{k-2} = \binom{k+l-4}{k-2}.$$

Recall that the set A is with no $(k-1)$-cups and that the set B' is with no k-cups.

Since the slope of any line segment with one end point in A and the other end point in B' is greater than any slope of the line segment between two points in B', any cup in S that contains points from both A and B' can contain only one point from B'. Therefore, the set S contains no k-cup. Similarly, the set S contains no l-cap.

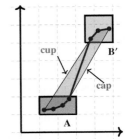

Hence the set S has the $(k,l)^+$-property. This completes the induction step. $\qquad\square$

We observe that Proposition 6.3.1 establishes that $f(k,l) > \binom{k+l-4}{k-2}$.

6.3.1 An Upper Bound

Already in 1935, Erdős and Szekeres established the exact value of $f(k,l)$, for any $k,l \geq 3$.

Theorem 6.3.1 ([28]). *If $k,l \geq 3$ then $f(k,l) = \binom{k+l-4}{k-2} + 1$.*

Proof. We prove the claim via induction on $k+l$.

Clearly, any three points in general position will make a three-cup or a three-cap.
On the other hand, $\binom{3+3-4}{3-2} + 1 = 2 + 1 = 3$.

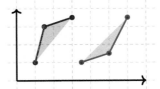

Therefore, $f(3,3) = \binom{3+3-4}{3-2} + 1$ and we have established the base of induction: if $k,l \geq 3$ are such that $k+l = 6$ then $f(k,l) = \binom{k+l-4}{k-2} + 1$.

Assume that $m \geq 6$ is such that, for any $u,v \geq 3$, if $u+v = m$ then $f(u,v) = \binom{u+v-4}{u-2} + 1$.

Let $k,l \geq 3$ be such that $k+l = m+1$. Then $(k-1)+l = k+(l-1) = m$ and, by our assumption, $f(k-1,l) = \binom{k+l-5}{k-3} + 1$ and $f(k,l-1) = \binom{k+l-5}{k-2} + 1$.

Since, by Theorem 6.2.2, $f(k,l) \leq f(k-1,l) + f(k,l-1) - 1 = \binom{k+l-4}{k-2} + 1$, and since $\binom{k+l-4}{k-2} < f(k,l)$, the induction step follows. $\qquad\square$

Clearly, Theorem 6.3.1, together with Theorem 6.2.2, establishes the following upper bound of $ES(n)$:

Corollary 6.3.1 ([28]). $ES(n) \le \binom{2n-4}{n-2} + 1$.

6.3.2 A Lower Bound

In 1960, Erdős and Szekeres provided a procedure that, starting with a one-element set, inductively generated a family of $n-1$ sets of points in the plane [29]. Besides two one-element sets, the family contains sets with the $(k, n-k)^+$-property, for $k \in [3, n-3]$. Their union, a set with 2^{n-2} points in general position, contains no convex n-gon and therefore establishes a lower bound of $ES(n)$.

In what follows, by using the main idea of the Erdős–Szekeres procedure, we obtain a relatively simple construction of a set with 2^{n-2} points in general position that contains no convex n-gon.

Before we start with our construction, we observe that if a set $S = \{(x_i, y_i) : 1 \le i \le \binom{k+l-4}{k-2}\}$ has the $(k, l)^+$-property then, for any $\varepsilon > 0$, the set $\varepsilon S = \{(\varepsilon x_i, \varepsilon y_i) : 1 \le i \le \binom{k+l-4}{k-2}\}$ also has the $(k, l)^+$-property. In addition, if the diameter of the set S, i.e. the largest distance between two points in S, equals $d > 0$, then the diameter of the set εS is εd.

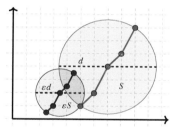

In other words, for any $k, l \ge 3$, there is a set with the $(k, l)^+$-property with an arbitrarily small diameter.

Recall that we have established that $ES(n) = 2^{n-2} + 1$ for $n \in \{4, 5, 6\}$. Hence, we suppose that $n \ge 7$.

Let a rectangular coordinate system be given.

Let $A = \{A_1, A_2, \ldots, A_{n-1}\}$ be a cap such that A_1 is its most left point and, for any $i \in [1, n-2]$, the slope of the line passing through A_i and A_{i+1} is negative. For example, A may be the set of vertices of a regular $4(n-1)$-polygon, centred at the origin, that lie in the first quadrant. For each $i \in [1, n-1]$, we construct a circle C_i with the centre at A_i and the diameter r. We choose $r > 0$ so that, for $i < j$, any point that belongs to C_j is to the right and below of any point that belongs to C_i.

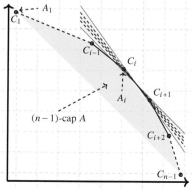

Hence the slope of any line that intersect two circles is negative. In addition, the diameter r is so small that any line in the plane intersects at most two of the circles $C_1, C_2, \ldots, C_{n-1}$.[3]

For every $i \in [2, n-2]$, let $R_{i+1,n-i+1}$ be a set with the $(i+1, n-i+1)^+$-property and with the diameter less than r. Inside of each circle C_i we place a set S_i, where $S_1 = \{A_1\}$, $S_{n-1} = \{A_{n-1}\}$, and, for $i \in \{2, \ldots, n-2\}$, S_i is a translate of $R_{i+1,n-i+1}$.

Let $S = \cup_{i=1}^{n-1} S_i$.

Theorem 6.3.2. *The set S is a set with 2^{n-2} points in general position that does not contain vertices of a convex n-gon.*

Proof. Since, for $i \neq j$, S_i and S_j are disjoint sets, it follows that $|S| = \sum_{i=1}^{n-1} |S_i|$. Recall that $|S_1| = 1 = \binom{n-2}{0}$, $|S_{n-1}| = 1 = \binom{n-2}{n-2}$, and observe that, for $i \in \{2, \ldots, n-2\}$, $|S_i| = |R_{i+1,n-i+1}| = \binom{n-2}{i-1}$. Therefore, $|S| = \sum_{i=0}^{n-2} \binom{n-2}{i} = 2^{n-2}$.

As a translate of a set of points in general position, each S_i, $i \in [2, n-2]$, is a set of points in general position. This together with the fact that any line with a negative slope can intersect the set S_i, $i \in [1, 2, n-1]$, in at most one point, and the fact that no line intersects more than two of $S_1, S_2, \ldots, S_{n-1}$, implies that the set S is a set of points in general position.

Let m be a positive integer and let $\Pi = P_1 P_2 \cdots P_m$ be a convex m-gon, oriented clockwise, with $P_i \in S$, for all $i \in [1, m]$. Let k be the least $i \in [1, n-2]$ among all those S_i's that contain a vertex of the polygon Π. Let P_1 be the vertex with the smallest x-coordinate in S_k.

Suppose that all vertices of the polygon Π belong to S_k. In general, Π may consist of an s-cup and a t-cap with the common endpoints P_1 and P_ℓ, $\ell < m$.

If this is the case, then $m = s + t - 2$. Recall that in S_k all cups are of the size at most k and all caps are of the size at most $n - k$. Hence, $m \leq k + (n - k) - 2 = n - 2$. If P_1 and P_m are the end points of a cup or a cap contained in S_k, then $m \leq \max\{k, n - k\} \leq n - 2$. Therefore no S_i, $i \in [2, n-2]$, contains a convex n-gon.

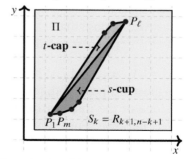

Suppose that there is $i \in [2, m]$ such that $P_i \notin S_k$. Let l be the largest among all $i \in [k+1, n-2]$ for which S_i contains a vertex of Π.

---

[3]We can use calculus to establish that such $r > 0$ exists, i.e. that r is so small that the value of the slope of any line that intersects both C_i and C_j, $i \neq j$, is as close to the slope of the line through A_i and A_j as we wish.

Under this assumption, the vertices of Π
contained in S_k do not form a cup in S_k.
Otherwise, as shown in the figure to the
right, the line segment $\overline{P_1 P_s}$ would be outside of the polygon Π. It follows that S_k
contains at most k (which is the size of the
largest possible cap in S_k, if $k > 1$) vertices
of the polygon Π.

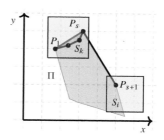

Similarly, the vertices of Π contained in S_l do not form a cap in S_l. Hence
S_l contains at most $n - l$ (which is the size of the largest possible cup in S_l, if
$l < n - 1$) vertices of the polygon Π.

Finally, suppose that $i \in [k + 1, l - 1]$ is such that S_i contains a vertex of Π.

Suppose that S_i contains more than one vertex of Π. Let $P_t \in S_i$ be such
that $P_{t-1} \notin S_i$. Let p_t be the line through P_{t-1} and P_t. Since Π is convex, all
of its vertices, except P_{t-1} and P_t are on the same side of the line p_t, say on
the right–hand side.

Let $s > t + 1$ be the smallest integer such
that the vertex P_s does not belong to S_i
but that the vertex $P_{s-1} \in S_i$. Consider the
side $\overline{P_t P_{t+1}}$ of the polygon Π. Since the
the point P_{s-1} is above this side, the whole
polygon is. But the slope of $\overline{P_t P_{t+1}}$ is positive and the slope of $\overline{P_t P_s}$ is negative,
which implies that the line segment $\overline{P_t P_s}$
is not contained in Π. This contradicts our
assumption that Π is convex.

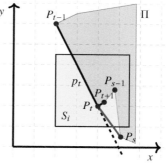

Therefore, for each $i \in [k + 1, l - 1]$, S_i contains at most one vertex of Π.

We have established that each vertex of Π is one of the most k vertices of
Π that belong to S_k or it is one of the most $n - l$ vertices of Π that belong to
S_l or it is the unique vertex that belongs to S_i for some $i \in [k + 1, l - 1]$. This
implies that $m \le k + (n - l) + ((l - 1) - (k + 1) + 1) = n - 1$.

It follows that the set S does not contain the vertices of a convex n-gon. $\quad\square$

Hence, the following lower bound:

Corollary 6.3.2. $ES(n) \ge |S| = 2^{n-2} + 1$.

Over the bridge of time:

In 1935, Erdős and Szekeres stated that it "is possible to give $\binom{2n-4}{n-2}$
points such that they contain neither convex nor concave points." [28]

In 1961, Erdős and Szekeres revisited this statement: "We have also stated, without proof, that exists a set $S_{k,l}$ of $f(k,l) = \binom{k+l-2}{k-1}$ points which contains no [k-cap] and no [l-cup]. We shall construct an explicit example of such a set." They also added: "We shall construct a set of 2^{n-2} points which contains no convex n-gon." [29]

It turned that that the Erdős–Szekeres construction "contained some minor inaccuracies, which were all corrected by Kalbfleisch [and Stanton]," as Erdős commented in 1975 [30].

James Kalbfleisch,1940–2017, and Ralph Stanton, 1923–2010, Canadian mathematicians, published the corrected construction in 1995, together with the note: "This paper was written in 1966 but, because of a series of coincidences, was not published at the time" [63].

6.4 PROGRESS ON THE CONJECTURE OF ERDŐS AND SZEKERES

"Truth is much too complicated to allow anything but approximations."

John von Neumann, a Hungarian–American
mathematician, physicist, and computer scientist

1903–1957

As we have demonstrated, already in 1935, Erdős and Szekeres knew that $2^{n-2} + 1 \leq ES(n) \leq \binom{2n-4}{n-2} + 1$ and commented that the upper bound was "too large."

One way to grasp the size of the above upper bound is to use Stirling's formula[4] $n! \approx \sqrt{2\pi n} \cdot (\frac{n}{e})^n$ to obtain that $\binom{2n-4}{n-2} + 1 \approx \frac{4^{n-2}}{\sqrt{\pi(n-2)}}$. We observe that, for any small positive number c and a large enough natural number n,

$$4^{n-2} > \frac{4^{n-2}}{\sqrt{\pi(n-2)}} > \frac{4^{n-2}}{(1 + \frac{c}{4-c})^{n-2}} = (4-c)^{n-2}.$$

This suggests that, for large $n \in \mathbb{N}$, $\binom{2n-4}{n-2} + 1$ is essentially as big as 4^{n-2}. In comparison with $2^{n-2} + 1$, the Erdős–Szekeres' upper bound *is* too large!

The upper bound of $ES(n)$ stayed the same for more than six decades. Then, in 1998, there were three improvements published in the same special issue of Discrete & Computational Geometry dedicated to the memory of Erdős.

---

[4]James Stirling, a Scottish mathematician, 1692–770. For a proof of Stirling's formula see, for example, [94].

Chung and Graham proved that $ES(n) \leq \binom{2n-4}{n-2}$.

Daniel Kleitman, an American mathematician, and Lior Pachtler, an Is-raeli–born computational biologist, established that $ES(n) \leq \binom{2n-4}{n-2} + 7 - 2n$.

Geza Tóth, a Hungarian mathematician, and Pavel Valtr, a Czech mathe-matician, proved that $ES(n) \leq \binom{2n-5}{n-2} + 2$.

Seven years later, in 2005, Tóth and Valtr made a further improvement by proving that $ES(n) \leq \binom{2n-5}{n-2} + 1$.

For proofs of all four results listed above and the related references, see [120].

From

$$\binom{2n-4}{n-2} = \frac{(2n-4)!}{((n-2)!)^2} = \frac{(2n-4)\cdot(2n-5)!}{(n-2)\cdot(n-3)!\cdot(n-2)!} = 2\cdot\binom{2n-5}{n-2}$$

we see that the upper bound established by Tóth and Valtr is approximately twice as good as the result of Erdős and Szekeres. Still, for large n, this does not change the magnitude of the upper bound.

Only in 2017, Andrew Suk, an American mathematician, was able to find a way to decrease the magnitude of the upper bound of $ES(n)$. Namely, Suk established that there is a large absolute constant n_0 such that $n \geq n_0$ implies that $ES(n) \leq 2^{n+6\sqrt[3]{n^2}\log_2 n}$ [113].

To illustrate the significance of Suk's result, we recall that from $\lim_{n\to\infty} \frac{6\sqrt[3]{n^2}\log_2 n}{n} = 0$ it follows that, for any $\varepsilon > 0$, there is n_ε such that $n \geq n_\varepsilon$ implies that $6\sqrt[3]{n^2}\log_2 n < n\varepsilon$. Therefore, for any $n \geq \max\{n_0, n_\varepsilon\}$, $ES(n) < 2^{(1+\varepsilon)n}$.

For example, thanks to Suk, we *know* that for all sufficiently large n we have that $ES(n) \leq 2^{1.000001n}$.

Having an open mind at the right time.

Suk summarized his result by saying "we nearly settle the Erdős–Szekeres conjecture [113]."

We observe that Suk obtained his celebrated result by showing the ex-istence of cups and caps of appropriate sizes in any set of $\lfloor 2^{n+6\sqrt[3]{n^2}\log_2 n} \rfloor$ points in general position. To do so, he was basically using objects intro-duced by Erdős and Szekeres in 1935, together with several related mathe-matical facts established over time. Because of this, one may call Suk's proof *elementary*, as Graham did in [55].

Still, Suk's proof is another example of a very human side of mathemat-ics: Everything may be in the front of our eyes for a long time and all the necessary tools may be at our reach, but a moment of human brilliance is needed to apply the right tools in the right order and in the right way to unveil the hidden gems.

In Paul Erdős's words:

"It is not enough to be in the right place at the right time. You should also have an open mind at the right time [107]."

6.5   EXERCISES

Exercise 6.1 (*Convex quadrilaterals*)**.** Show that, given five points in the plane in general position, the number of convex quadrilaterals formed by these points is odd.

Exercise 6.2 (*Convex hexagon*)**.** Find ten points in the plane in general position that do not contain a convex hexagon.

Exercise 6.3 (*Empty convex quadrilateral*)**.** Prove that any five points in the plane in general position contain an *empty* convex quadrilateral, i.e. a convex quadrilateral that does not contain in its interior the remaining point. Is this true if you take more than five points in general position?

Exercise 6.4 (*Tarsi's solution to the Happy End Problem*)**.** While he was a student, Michael Tarsi, an Israeli mathematician, solved the Happy End Problem in the following way:

1. For $n \geq 4$, let S be a set of points in the plane in general position of the size $m = R(3; n, n)$. Enumerate the elements of the set S by numbers from 1 to m.

2. Colour three-element subsets of S as follows:

   - Colour the set $\{i, j, k\}$, $i < j < k$, red if you travel from i to j to k in a clockwise direction.
   - Colour the set $\{i, j, k\}$, $i < j < k$, blue if you travel from i to j to k in a counterclockwise direction.

Complete Tarsi's proof by applying Ramsey's theorem!

Exercise 6.5 (*The number $f(5,5)$*)**.** Take the following facts as given: (a) $f(k, l) = f(l, k)$; (b) From $f(k, l) \leq \binom{k+l-4}{k-2} + 1$ it follows that $f(4, 4) \leq 7$, $f(5, 4) \leq 11$, and $f(5, 5) \leq 21$; and (c) $f(k, 3) = f(3, k) = k$.

Prove that $f(5, 5) = 21$.

Exercise 6.6 (*Vertical compression and homothety of sets*). Show that, for any $k, l \geq 3$ and any $\varepsilon, r > 0$, there is a set $F_{k,l}$ of $\binom{k+l-4}{k+2}$ points in general position, with no k-cups or l-caps, and with the following two properties:

1. Each line that contains two points of $F_{k,l}$ is with the slope whose absolute value is less than ε.

2. The diameter of $F_{k,l}$ is less than or equal to r.

Exercise 6.7 (*Lovász's construction*). In [77], László Lovász, a Hungarian mathematician, constructed a set with 2^{n-2} points in general position that contains no convex n-gon in the following way:

Let, for $n \geq 4$, P be a regular $(4n - 4)$-gon with centre at the origin and let $V_0, \ldots, V_{n-2}$ be its vertices in the segment of $\pm 45^0$ around the positive half of the x-axis labelled clockwise (so that V_0 is the highest point).

Let $S_0 = \{V_0\}$, $S_{n-2} = \{V_{n-2}\}$, and let, for $i \in [1, n-3]$, S_i be a set of $\binom{n-2}{i}$ points in general position, with no $(n-i)$-cups or $(i+2)$-caps, contained in a circle with the centre at V_i and radius r, and such that each line that contains two points of S_i is with the slope whose absolute value is less than 1. (See Exercise 6.6.)

Prove that, for a sufficiently small $r > 0$, the set $S = \cup_{n=0}^{n-2} S_i$ is a set with 2^{n-2} points in general position that contains no convex n-gon.

Exercise 6.8 (*Geometry of the Moser spindle*). In 1961, Leo and William Moser introduced a geometric object consisting of seven vertices and eleven line segments of the unit length.

This object is now known as the Moser spindle [60, 85]. See the figure to the right. Let A be the vertex of degree four and let C and D be two vertices that are not adjacent to A. Let α be the measure of $\angle DAC$ and let x be the measure of the angle $\angle GAF$ in degrees.

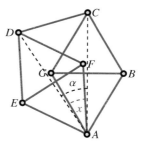

1. Determine $|\overline{AC}|$ and $|\overline{AD}|$.

2. Determine the value of α and the value of x.

3. Construct the Moser spindle by starting with drawing two concentric circles, one with radius 1 and the other with radius $|\overline{AC}|$.

Exercise 6.9 (*Moser spindle in action*). In the May 1961 issue of the Canadian Mathematical Bulletin [85], Leo and William Moser posted their solution to the following problem:

1. Prove that every set of six points in the plane can be coloured in three colours in such a way that no two points are a unit distance apart have the same colour.

2. Show that in (1) six cannot be replaced by seven.

The problem was proposed by the Moser brothers in the late 1950s.

In other words, Leo and William Moser established that χ, the chromatic number of the plane,[5] is greater than or equal to 4. See, for example, [60, 61].

In this exercise, we follow Leo and William Moser's solution to the problem above.

Let two points which are a unit distance apart be called *friends*, otherwise they are *strangers*. If a finite set of points can be coloured by k colours, so that no pair of friends have the same colour, we say that this set permits a *proper k-colouring*. In the rest of this exercise we assume that any set of four or more points is a subset of a plane.

1. Prove that if the points in the plane are coloured by one of the two given colours, then there must exist two points that are at a unit distance and coloured by the same colour.

2. Show that four points in the plane cannot be friends to each other and that two points cannot have three common friends.

3. Show that any set of four points permits a proper three-colouring.

4. Show that any set of five points permits a proper three-colouring.

5. Show that any set of six points permits a proper three-colouring.

6. Show that there is a set of seven points that does not permit a proper three-colouring.

Exercise 6.10 (*A taste of de Gray's proof*). In 2018, Aubrey de Grey, an English author and biomedical gerontologist, by constructing a unit distance graph with 20,425 vertices and with its chromatic number greater than 4, established that the chromatic number of the plane χ is greater than or equal to 5 [21].

---

[5]The chromatic number of the plane is the smallest number of colours sufficient for colouring the plane in such a way that no two points of the same colour are a unit distance apart.

de Grey's proof was a combination of new insights into some of the well-known facts and techniques and a computer-assisted mathematical proof. In de Grey's words:

> In seeking graphs that can serve as [a unit distance graph with chromatic number greater than four] in our construction, we focus on graphs that contain a high density of Moser spindles. The motivation for exploring such graphs is that a spindle contains two pairs of vertices distance $\sqrt{3}$ apart, and these pairs cannot both be monochromatic. Intuitively, therefore, a graph containing a high density of interlocking spindles might be constrained to have its monochromatic $\sqrt{3}$-apart vertex pairs distributed rather uniformly (in some sense) in any four-colouring. Since such graphs typically also contain regular hexagons of side-length 1, one might be optimistic that they could contain some such hexagon that does not contain a monochromatic triple in any four-colouring of the overall graph, since such a triple is always an equilateral triangle of edge $\sqrt{3}$ and thus constitutes a locally high density, i.e. a departure from the aforementioned uniformity, of monochromatic $\sqrt{3}$-apart vertex pairs.

1. Confirm de Grey's observation that "a spindle contains two pairs of vertices distance $\sqrt{3}$ apart, and these pairs cannot both be monochromatic" in a proper k-colouring of the Moser spindle, $k > 3$.

2. Use your graphing tool to rotate clockwise the Moser spindle through α (established in Exercise 6.8) about the point A to obtain another Moser spindle. Call the newly obtained graph, i.e. the graph that consists of the original Moser spindle and its image obtained by the rotation, MS_2. What do you observe? Is this a unit distance graph? Does MS_2 contain a hexagon, not necessarily regular, "of side-length 1?"

3. Find a proper four-colouring of MS_2. Do this in two ways, by hand and by writing a program that would search for proper four-colourings of MS_2.

4. Augment the drawing of MS_2 by rotating clockwise the original Moser spindle by $\frac{\alpha}{2}$ about the point A. Find a proper four-colouring of the newly obtained graph on 16 vertices.

Exercise 6.11 (*Density and chromatic number*). Hallard T. Croft, an English mathematician, attributed the "density version" of the problem of finding the chromatic number of the plane to Leo Moser: What is $m_1(\mathbb{R}^2)$, the maximum density of a measurable set in the plane that does not contain a unit distance pair [19]?

In more casual terms, the question is to determine the size (in terms of the portion of the plane) that a set that avoids the unit distance cannot exceed.

At the time of writing this text, the best upper bound of $m_1(\mathbb{R}^2)$ is 0.25688. Erdős said that "it seems very likely" that $m_1(\mathbb{R}^2) < 0.25$.

In 1967, Croft established that $0.2293 < m_1(\mathbb{R}^2) \leq \frac{2}{7} = 0.2857$ and commented that "the bounds are surprisingly close."

In this exercise we will follow Croft's argument that $m_1(\mathbb{R}^2) > 0.2293$. Interestingly enough this is still the best-known lower bound of $m_1(\mathbb{R}^2)$.

Start by covering the plane with an infinite equilateral triangular lattice. The length of the side of the triangles in the lattice will be determined by the following construction.

Say that $\triangle XYZ$ is one of the triangles in the lattice. We construct three mutually congruent "lumps" in $\triangle XYZ$ in the following way:

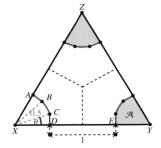

- Choose an angle $\theta \in \left(0, \frac{\pi}{6}\right)$.

- Let B and C be points inside of $\triangle XYZ$ such that $\overline{XB} = \overline{XC} = \frac{1}{2}$ and that $\angle(YXC) = \angle(ZXB) = \theta$.

- Draw the arc BC on the circle with the centre at X and radius equal to $\frac{1}{2}$.

- Let A be the point on the line segment $\overline{XZ}$ such that $\overline{AB} \perp \overline{XZ}$.

- Let D be the point on the line segment $\overline{XY}$ such that $\overline{CD} \perp \overline{XY}$.

- Call the interior of the region in the plane bounded by the line segments $\overline{XA}$, $\overline{XD}$, $\overline{AB}$, $\overline{CD}$, and the arch BC, a "lump" centred at X with the angle θ.

- Do the same construction for the vertices Y and Z. Keep the angle θ same as above to obtain two more congruent lumps in $\triangle XYZ$.

- Choose $|\overline{XY}|$, the length of the side of the equilateral triangle so that the distance between the corresponding points (obtained in this construction) on the line segment $\overline{XY}$ equals to 1.

- Do the same construction for all triangles in the lattice.

Let $S = S(\theta)$ be the set of all points in the plane that belong to one of the lumps obtained by the above construction for some (fixed) θ.

1. What is the length of the line segment $\overline{XY}$?

2. Show that if $M, N \in S$ then $|\overline{MN}| \neq 1$.

3. Find the area $\mathcal{A} = \mathcal{A}(\theta)$ of a lump.

4. Determine the area $A_\triangle = A_\triangle(\theta)$ of a triangle in the lattice.

5. Find the density $\delta(S) = \delta(S(\theta))$ of the set $S = S(\theta)$.

6. Use your knowledge of calculus to find the maximum value of $\delta(S(\theta))$. Use technology to approximate the maximum value.

7. Are you convinced that $m_1(\mathbb{R}^2) > 0.2293$? Why yes or why not?

Exercise 6.12 (*Sylvester–Gallai theorem*). In 1990, Canadian mathematicians Peter Borwein, 1953–2020, and William Moser, in their paper *A survey of Sylvester's problem and its generalizations* [12], wrote:

> Let a finite set of points in the plane have the property that the line through any two of them passes through a third point of the set. Must all the points lie on one line? Almost a century ago Sylvester[6] (1893) posed this disarmingly simple question. No solution was offered at that time and the problem seemed to have been forgotten. Forty years later it resurfaced as a conjecture by Erdős: If a finite set of points in the plane are not all on one line then there is a line through exactly two of the points. In a recent reminiscence Erdős wrote: "I expected it to be easy but to my great surprise and disappointment I could not find a proof. I told this problem to Gallai who very soon found an ingenious proof." In 1943 Erdős proposed the problem in the American Mathematical Monthly, still unaware that it had been asked fifty years earlier, and the following year Gallai's solution appeared in print. Since then there has appeared a substantial literature on the problem and its generalizations.

In what follows, we slightly modify the argument due to L. M. Kelly.[7] See, for example, [1].

Let $\mathcal{P} = \{A_1, A_2, \ldots, A_n\}$ be a set of non-collinear points in the plane. Let $\mathcal{L}$ be the set of all lines in the plane that pass through at least two points that belong to $\mathcal{P}$.

---

[6]James Joseph Sylvester, an English mathematician, 1814–1897.
[7]Leroy Milton Kelly, an American mathematician, 1914–2002.

1. Conclude that the set $\mathcal{L}$ is finite, i.e. conclude that $\mathcal{L} = \{\ell_1, \ell_1, \ldots, \ell_m\}$, for some $m \geq 2$.

2. For each $i \in [1, m]$, let $\mathcal{P}_i \subset \mathcal{P}$ be the set of all points in $\mathcal{P}$ that do not lie on ℓ_i. Conclude that, for each $i \in [1, m]$, $\mathcal{P}_i \neq \emptyset$.

3. For each $i \in [1, m]$ and each $A \in \mathcal{P}_i$, let $d_{i,A} > 0$ be the distance between the point A and the line ℓ_i.

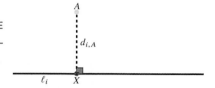

Conclude that there are $i_0 \in [1, m]$ and $A_{i_0} \in \mathcal{P}_{i_0}$ such that $d_{i_0, A_{i_0}} = d = \min\left(\cup_{i=1}^{m}\{d_{i,A} : A \in \mathcal{P}_i\}\right) > 0.$

4. Suppose that $i \in [1, m]$ is such that the line ℓ_i contains three points $K, M, N \in \mathcal{P}$. Let $A \in \mathcal{P}_i$ and let $X \in \ell_i$ be such that $\overline{AX} \perp \ell_i$. Prove that $d_{i,A} > d$ by considering two cases, $X \in \{K, M, N\}$ and $X \notin \{K, M, N\}$.

5. Conclude that there is a line in $\mathcal{L}$ passing through exactly two points that belong to $\mathcal{P}$.

CHAPTER 7

Solutions

This chapter contains solutions to all exercises.

7.1 CHAPTER 2: RAMSEY'S THEOREM

Exercise 2.1 If there are 11 pigeons (picked shoes) sitting in 10 pigeon-holes (one for each pair of shoes), by the pigeonhole principle, at least one pigeonhole has 2 pigeons (so a pair of shoes). □

Exercise 2.2 Let a colouring $f : [1,82] \times [1,4] \to \{\bullet, \blacksquare, \blacktriangle\}$ be given. For each $i \in [1,82]$ we define g_i, a colouring of the set $[1,4]$, by $g_i(1) = f(i,1)$, $g_i(2) = f(i,2)$, $g_i(3) = f(i,3)$, $g_i(4) = f(i,4)$.

Let P be the the the set of all functions $g : [1,4] \to \{\bullet, \blacksquare, \blacktriangle\}$, i.e. let P be the set of all three-colourings of the set $[1,4]$.

Observe that $|P| = 3^4 = 81$. In addition, observe that, by the pigeonhole principle, for each $g \in P$ there are $c, d \in [1,4]$, $c < d$, such that $g(c) = g(d)$.

Let the elements of the set P be pigeonholes and let the elements of the set $\{g_i : i \in [1,82]\}$ be pigeons.

By the pigeonhole principle, at least one of 81 pigeonholes contains at least two of 82 pigeons. In other words, there is $g \in P$ and $a, b \in [1,82]$, $a < b$, such that $g = g_a$ and $g = g_b$. This means that, for each $j \in [1,4]$, $g_a(j) = g_b(j)$.

Recall that there are $c, d \in [1,4]$, $c < d$, such that $g_a(c) = g_a(d) = g_b(c) = g_b(d)$.

By definition this means that $f(a,c) = f(a,d) = f(b,c) = f(b,d)$.

Therefore the rectangle with vertices (a,c), (a,d), (b,c), and (b,d) has all its vertices in the same colour. □

Exercise 2.3 Since there are 144 points in the grid, by the pigeonhole principle, there are at least 48 points of the same colour, say red. Let R be the set of red points.

DOI: 10.1201/9781003286370-7

For each $k \in [0, 12]$, let a_k be the number of columns with exactly k red points. Then, $\sum_{k=0}^{12} a_k = 12$ and $\sum_{k=0}^{12} k a_k = |R| \geq 48$.

Observe that, by the pigeonhole principle, there is $k \in [4, 12]$ such that $a_k > 0$. Also, recall that for any integer k, $(k-3)^2 = (k^2 - k) - (5k - 9) \geq 0$ implies that $k^2 - k \geq 5k - 9$. In particular, if $k \geq 4$ then $k^2 - k > 5k - 9$.

Next we count the number of pairs of red points that belong to the same column:

$$\sum_{k=2}^{12} \binom{k}{2} a_k = \frac{1}{2} \cdot \sum_{k=0}^{12} (k^2 - k) a_k > \frac{1}{2} \cdot \sum_{k=0}^{12} (5k - 9) a_k = \frac{5|R| - 9 \cdot 12}{2} \geq 66 = \binom{12}{2}.$$

By the pigeonhole principle, there are two different columns with red pairs in the same (vertical) position. Therefore there is a red rectangle. □

Exercise 2.4 Choose $\{a_1, \ldots, a_{n+1}\}$ from $[1, 2n]$, so that they are distinct. For each $i \in [1, n+1]$, we may write $a_i = 2^{b_i} q_i$, where q_i is an odd number contained in $[1, 2n]$. Consider the set $\{q_1, q_2, \ldots, q_{n+1}\}$. Since there are only n odd numbers in the interval $[1, 2n]$, by the pigeonhole principle, for some $i \neq j$, $q_i = q_j$. Let $q = q_i = q_j$. Then $a_i = 2^{b_i} q$ and $a_j = 2^{b_j} q$.

Since $a_i \neq a_j$, we have that either $b_i > b_j$ or $b_j > b_i$. In the former case, we have that $a_j | a_i$ and in the latter case, $a_i | a_j$, as required. □

Exercise 2.5 Given numbers $1, 2, \ldots, 12$ placed around a circle in some order, partition the circle into four disjoint sections, each containing three consecutive (in the order that they are placed around the circle) numbers. Let A_1, A_2, A_3, and A_4 be the four sets of consecutive numbers created by this partition.

Let $a_j = \sum_{a \in A_j} a$ for $j = 1, 2, 3, 4$. We know that $a_1 + a_2 + a_3 + a_4 = \sum_{b=1}^{12} b = 78$.

By the generalized pigeonhole principle, some a_i for $i \in [1, 4]$ has the property that $a_i \geq \left\lceil \frac{78}{4} \right\rceil = \lceil 19.5 \rfloor = 19$, as required. □

Exercise 2.6 Let $a_1, a_2, \ldots, a_{50}$ be a sequence of positive integers. For each $i \in [1, 50]$, let m_i be the length of the longest non-decreasing subsequence starting at and including a_i. This means that there are integers $j_1, j_2, \ldots, j_{m_i} \in [1, 50]$ such that $i = j_1 \leq j_2 \leq \cdots \leq j_{m_i}$ and $a_i = a_{j_1} \leq a_{j_2} \leq \cdots \leq a_{j_{m_i}}$, but for any $s \in \{j_{m_i} + 1, \ldots, 50\}$, $a_{j_{m_i}} > a_s$.

If, for some i, $m_i \geq 8$, then our observation has been supported. Hence, let us suppose that, for all $i \in [1, 50]$, $m_i \leq 7$.

Next, let the elements of the set $[1, 7]$ be the pigeonholes and let the elements of the set $\{m_i : i \in [1, 50]\}$ be the pigeons. We put the pigeon m_i in the pigeonhole j if and only if $m_i = j$.

Observe that one of the pigeonholes contains at least eight pigeons. Otherwise, the number of pigeons would be at most $7 \cdot 7 = 49$.

Say that $i_1, i_2, \ldots, i_8 \in [1, 50]$ and $m \in [1, 7]$ are such that $i_1 \le i_2 \le \cdots \le i_8$ and $m_{i_1} = m_{i_2} = \cdots = m_{i_8} = m$.

What can we tell about the subsequence $a_{i_1}, a_{i_2}, \ldots, a_{i_8}$?

Recall that, for each $j \in [1, 8]$, $m_{i_j} = m$ represents the length of the longest non-decreasing subsequence starting at a_{i_j}.

Let $j, k \in [1, 8]$, $j > k$. Observe that, under our assumptions, $a_{i_j} > a_{i_k}$. Otherwise, we will have a non-decreasing sequence starting at a_{i_j} and of length $m + 1$. (In this scenario, a_{i_j} would be followed by an m-term non-decreasing subsequence determined by a_{i_k}.)

Therefore $a_{i_1} > a_{i_2} > \cdots > a_{i_8}$ and we have found an eight-term non-increasing subsequence. ◻

Exercise 2.7 Suppose that there are n guests at the party. Each guest can be a friend with k other guests, where $k \in [0, n-1]$. Observe that it is impossible to have two guests so that one of them has zero friends and the other has $n - 1$ friends. (Why?)

Let the party attendees be pigeons. We put each guest/pigeon into the pigeonhole that is labelled by the number of friends that the guest has. Since there are at most $n - 1$ pigeonholes, by the pigeonhole principle there must be a pigeonhole with at least two pigeons. In other words, there are at least two guests with the same number of friends. ◻

Exercise 2.8 Observe that if $k + n \in B$ then $a = b = k + n$ implies $a + b = 2k + 2n$.

Suppose that $k + n \notin B$. Consider n pigeonholes $\{k, k + 2n\}, \{k + 1, k + 2n - 1\}, \ldots, \{k + i, k + 2n - i\}, \ldots, \{k + n - 1, k + n + 1\}$. Take the elements of the set B to be pigeons. Since there are at least $n + 1$ pigeons (and none of them is equal to $k + n$) there is a pigeonhole that contains two pigeons, say the pigeonhole $\{k + j, k + 2n - j\}$, for some $j \in [1, n - 1]$.

Hence, $\{k + j, k + 2n - j\} \subset B$ and $(k + j) + (k + 2n - j) = 2k + 2n$. ◻

Exercise 2.9 Take any subset $B \subset A$ that contains at least 2024 elements. Then, by the pigeonhole principle, there are $a, b \in B$, $a < b$, such that a and b have the same remainder when divided by 2023. In other words, there are $p, q \in \mathbb{N}$, $p < q$, and $r \in [0, 2022]$ such that $a = 2023p + r$ and $b = 2023q + r$. But this implies that $b - a$ is divisible by 2023: $b - a = (2023q + r) - (2023p + r) = 2023(q - p)$.

On the other hand, $b - a = \underbrace{111\ldots1}_{s} - \underbrace{111\ldots1}_{t} = \underbrace{1\ldots1}_{s-t}\underbrace{0\ldots0}_{t} = 10^t \cdot c$, where $c = \underbrace{1\ldots1}_{s-t} \in A$ and $s, t \in \mathbb{N}$. Therefore, $b - a = 2023(q - p) = 10^t \cdot c$.

Since 2023 is not divisible by neither 2 nor 5, it must be a factor of $c \in A$. ◻

Exercise 2.10 Let $m \geq 3$. We represent the interval $[0, 1]$ as the union of m intervals of length $\frac{1}{m}$, $[0, 1] = \left[0, \frac{1}{m}\right] \cup \left[\frac{1}{m}, \frac{2}{m}\right] \cup \cdots \cup \left[\frac{m-1}{m}, 1\right]$.

By the pigeonhole principle, there are $p, q \in \mathbb{N}$, $p < q$, and $i \in \{0, 1, \ldots, m-1\}$ such that $\{p\alpha\}, \{q\alpha\} \in \left(\frac{i}{m}, \frac{i+1}{m}\right)$.

From $0 < (q - p)\alpha = \lfloor q\alpha \rfloor - \lfloor p\alpha \rfloor + \{q\alpha\} - \{p\alpha\}$, it follows that $\lfloor (q - p)\alpha \rfloor = \lfloor q\alpha \rfloor - \lfloor p\alpha \rfloor$, if $\{q\alpha\} - \{p\alpha\} > 0$, and $\lfloor (q-p)\alpha \rfloor = \lfloor q\alpha \rfloor - \lfloor p\alpha \rfloor - 1$, if $\{q\alpha\} - \{p\alpha\} < 0$. In the former case, $\{(q - p)\alpha\} = \{q\alpha\} - \{p\alpha\} \in \left(0, \frac{1}{m}\right)$, and in the latter case $\{(q - p)\alpha\} = 1 + \{q\alpha\} - \{p\alpha\} \in \left(\frac{m-1}{m}, 1\right)$.

Assume that $\{(q - p)\alpha\} \in \left(0, \frac{1}{m}\right)$. Then, for any $j \in [2, m]$, because $\{(q-p)\alpha\}$ cannot "jump" over an interval of length $\frac{1}{m}$, there is $r \in \mathbb{N}$ such that $r\{(q-p)\alpha\} \in \left(\frac{j-1}{m}, \frac{j}{m}\right)$. But then, $r(q - p)\alpha = r\lfloor (q-p)\alpha \rfloor + r\{(q-p)\alpha\}$ implies that $\{r(q - p)\alpha\} = r\{(q-p)\alpha\} \in \left(\frac{j-1}{m}, \frac{j}{m}\right)$.

If $\{(q-p)\alpha\} = 1 + \{q\alpha\} - \{p\alpha\} \in \left(\frac{m-1}{m}, 1\right)$, we set $t = \{p\alpha\} - \{q\alpha\} \in \left(0, \frac{1}{m}\right)$. As above, for any $j \in [1, m-1]$, there is $r \in \mathbb{N}$, such that $r(q - p)\alpha = r\lfloor (q-p)\alpha \rfloor + r - 1 + (1 - rt)$, with $1 - rt \in \left(\frac{j}{m}, \frac{j+1}{m}\right)$.

Therefore, F_α is dense in $[0,1]$.   □

Exercise 2.11

    a. Observe that for $r = 1$, $n = 2$, and $\mu = k$, Ramsey's theorem becomes:

> There is m_0 such that, for any $m \geq m_0$, if the elements of a set Γ_m (i.e. a set with m elements) are divided in any manner into k mutually exclusive classes C_i, $i = 1, 2, \ldots, k$, then Γ_m must contain a sub-class Δ_2 (i.e. a set with 2 elements) such that both elements of Δ_2 belong to the same C_i.

This is the claim of the pigeonhole principle. See Theorem 2.1.1. Note that $m_0 = k + 1$. We think about the elements of the set Γ_m as pigeons and about the classes C_i, $i = 1, 2, \ldots, k$, as pigeonholes.

    b. Observe that for $r = 1$, $n = N \geq 2$, and $\mu = k$, Ramsey's theorem becomes:

> There is m_0 such that, for any $m \geq m_0$, if the elements of a set Γ_m (i.e. a set with m elements) are divided in any manner into k mutually exclusive classes C_i, $i = 1, 2, \ldots, k$, then Γ_m must contain a sub-class Δ_N (i.e. a set with N elements) such that all elements of Δ_N belong to the same C_i.

If we take $m_0 = (N - 1)k + 1$, and if we think about the elements of the set Γ_m as pigeons and the classes C_i, $i = 1, 2, \ldots, k$, as pigeonholes, the above statement becomes the generalized pigeonhole principle. See Theorem 2.1.2. □

Exercise 2.12 Suppose that a graph G with at least six vertices is given. Colour all edges of the graph G red. Next, draw all missing edges and colour them blue. □

Exercise 2.13 Observe that there are 20 different triangles in K_6. Colour the edges of K_6 with red and blue and say that a triangle in K_6 is two-coloured if it is not monochromatic. By the pigeonhole principle, each two-coloured triangle contains two two-coloured angles, i.e. two angles with sides of different colours. Hence, the number of two-coloured triangles is equal to one-half of the number of two-coloured angles. Next we count the largest possible number of two-coloured angles.

Let A be a vertex in K_6. Observe that A may be the vertex of at most six differ- ent two-coloured angles. Since there are six vertices in K_6, the number of two-coloured angles is at most 36.

Hence, there are at most eighteen two-coloured triangles, i.e. there are at least two monochromatic triangles. □

Exercise 2.14

1. Suppose that there are $x, y \in [2, 16]$, $x < y$, such that K_4 with the vertices $0, 1, x, y$ is C-monochromatic.

   Since $1 - 0 = 1$, by definition, $C(\langle 0, 1 \rangle) = \bullet$.

   Since we have assumed that K_4 is C-monochromatic, from $C(\langle 0, x \rangle) = C(\langle 0, y \rangle) = \bullet$, it follows that $\{x, y\} \subset \{2, 4, 8, 9, 13, 15, 16\}$ and, from $C(\langle 1, x \rangle) = C(\langle 1, y \rangle) = \bullet$, it follows that $\{x, y\} \subset \{2, 3, 5, 9, 10, 14, 16\}$. Hence $\{x, y\} \subseteq \{2, 4, 8, 9, 13, 15, 16\} \cap \{2, 3, 5, 9, 10, 14, 16\} = \{2, 9, 16\}$. But, $C(\langle 2, 9 \rangle) = C(\langle 2, 16 \rangle) = C(\langle 9, 16 \rangle) = ■$, which contradicts our as- sumption that K_4 with the vertices $0, 1, x, y$ is C-monochromatic. There- fore, any K_4 that contains the edge $\langle 0, 1 \rangle$ is not C-monochromatic.

2. From $y - x = (y - x) - 0$, by definition of the colouring C, it follows that $C(\langle x, y \rangle) = C(\langle 0, y - x \rangle)$. For $a, b \in \{x, y, z, w\}$, $a < b$, the mapping $\langle a, b \rangle \mapsto \langle a - x, b - x \rangle$ establishes a 1–1 correspondence between the edges of K_4 with vertices x, y, z, w and the edges of K_4 with vertices $0, y - x, z - x, w - x$. In addition, $b - a = (b - x) - (a - x)$ implies $C(\langle a, b \rangle) =$

$C(\langle a - x, b - x \rangle)$, which implies that the corresponding edges are of the same colour.

3. (a)   i. From $C(\langle 0, x \rangle) = C(\langle 0, y \rangle) = C(\langle 0, z \rangle) = \bullet$, by definition of the colouring C, it follows that $\{x, y, z\} \subset \{2, 4, 8, 9, 13, 15, 16\}$. Since, $2 \le x < y < z$, we conclude that $x \in \{2, 4, 8, 9, 13\}$, $y \in \{4, 8, 9, 13, 15\}$, and $z \in \{8, 9, 13, 15, 16\}$.

   ii. Say $x = 2$. If $y = 4, z = 15$, then $C(\langle 2, 4 \rangle) = C(\langle 2, 15 \rangle) = \bullet$, but $C(\langle 4, 15 \rangle) = \blacksquare$. Otherwise, for $x = 2$ and $y > 4$, and for $x \in \{9, 13\}$, there is only one choice for $y \in \{4, 8, 9, 13, 15\}$ so that $C(\langle x, y \rangle) = \bullet$. Therefore, for $z > y$, $C(\langle x, z \rangle) = \blacksquare$. It follows that, in this case, K_4 is not monochromatic.

   iii. Observe that if $x \in \{4, 8\}$ only $x = 4 < y = 8 < z = 13$ and $x = 8 < y = 9 < z = 16$ imply $C(\langle x, y \rangle) = C(\langle x, z \rangle) = \bullet$. But in both cases $C(\langle y, z \rangle) = \blacksquare$.

   It follows that, in this case, K_4 is not monochromatic.

   (b)   i. From $C(\langle 0, x \rangle) = C(\langle 0, y \rangle) = C(\langle 0, z \rangle) = \blacksquare$, by definition of the colouring C, it follows that $\{x, y, z\} \subset \{3, 5, 6, 7, 10, 11, 12, 14\}$. Since, $3 \le x < y < z$, we conclude that $x \in \{3, 5, 6, 7, 10, 11\}$, $y \in \{5, 6, 7, 10, 11, 12\}$ and $z \in \{6, 7, 10, 11, 12, 14\}$.

   ii. Observe that if $x = 10$ then $y = 11$ or $y = 12$ and if $x = 11$ then $y = 12$. In each of these cases $C(\langle x, y \rangle) = \bullet$. It follows that, in this case, K_4 is not monochromatic.

   iii. If $x \in \{3, 5, 6, 7\}$ and $C(\langle x, y \rangle) = C(\langle x, z \rangle) = \blacksquare$ then the following cases are possible:

   - $x = 3 < y = 6 < z = 10$, $x = 3 < y = 6 < z = 14$, $x = 3 < y = 10 < z = 14$;
   - $x = 5 < y = 10 < z = 11$, $x = 5 < y = 10 < z = 12$, $x = 5 < y = 11 < z = 12$;
   - $x = 6 < y = 11 < z = 12$; and
   - $x = 7 < y = 10 < z = 12$, $x = 7 < y = 10 < z = 14$, $x = 7 < y = 12 < z = 14$.

   But in each of these cases $C(\langle y, z \rangle) = \bullet$. It follows that, in this case, K_4 is not monochromatic.

Note: Observe that $[0, 16]$ is the set of residues modulo 17. Recall that, since 17 is a prime number, the set $\{1, 2, \ldots, 16\}$, under the multiplication modulo 17, is a group, commonly denoted by $\mathbb{Z}_{17}^*$, with 1 as its unit/neutral element. The set $\{1, 2, 4, 8, 9, 13, 15, 16\}$, used in the definition of the colouring C, is the set of quadratic residues in $\mathbb{Z}_{17}^*$.

This particular colouring was discovered by Greenwood and Gleason in 1955. For their original argument see [52]. □

Exercise 2.15 By definition, $R(4,3) = 9$ means that any two-colouring of K_9 yields a monochromatic K_4 in the first colour OR a monochromatic K_3 in the second colours. So the colouring of K_{17} with no monochromatic K_4 will have many monochromatic complete graphs K_3. No contradiction! □

Exercise 2.16 Consider a three-edge-colouring c of K_{17} with colours c_1, c_2, c_3.

Fix x, a vertex in K_{17}. The vertex x is incident to 16 edges, coloured with one of the three colours. By the generalized pigeonhole principle, we have that there is a colour c_i, for some $i \in \{1, 2, 3\}$, such that x is incident to at least six edges coloured by c_i.

Let $Y_i = \{y_1, \ldots, y_6\}$ be a set of neighbours of the vertex x such that $c(\{x, y_j\}) = c_i$, for $j = 1, 2, \ldots, 6$. If any edge joining two vertices in Y_i is also coloured by colour c_i, then they form a monochromatic K_3 with x as its third vertex. Otherwise, every edge joining vertices of Y_i is not coloured by colour c_i. Then, the vertices of Y_i induce a subgraph of K_{17}, that is a copy of K_6, edge-coloured by two colours. By Ramsey's theorem, every two-colouring of K_6 contains a monochromatic K_3, and the result follows. □

Exercise 2.17 Our first step is to establish that $R_r(3) \le r R_{r-1}(3)$.

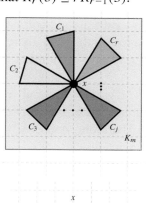

Let $r \ge 2$ and let $m = r R_{r-1}(3)$. We r-colour the edges of K_m, and fix a vertex x. For $j \in [1, r]$, let C_j be the set of all vertices a such that the edge $\{x, a\}$ is coloured by the colour i.

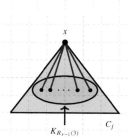

Use the pigeonhole principle to conclude that if the edges of K_m are r-coloured then a fixed vertex x is incident to at least $R_{r-1}(3)$ edges that are of the same colour, say blue.

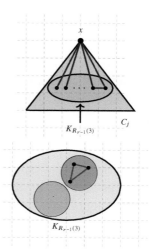

Case 1: If $K_{R_{r-1}(3)}$ contains a blue edge then a monochromatic (blue) K_3 appears.

Case 2: If $K_{R_{r-1}(3)}$ does not contain any blue edges, i.e. if at most $r-1$ colours are used, then, by definition of $R_{r-1}(3)$, there must be a monochromatic triangle there.

In both cases, the r-colouring of K_m contains a monochromatic triangle, so $R_r(3) \leq rR_{r-1}(3)$. Recall that $R_2(3) = R(3,3) = 6 = 3 \cdot 2!$ and observe that, if $R_{r-1}(3) \leq 3(r-1)!$, then $R_r(3) \leq rR_{r-1}(3) \leq r \cdot 3(r-1)! = 3r!$ □

Exercise 2.18 Take, for example, $X = \{2^k : k \in \mathbb{N}\}$. □

Exercise 2.19 Let $\mathcal{A} = \{\{x, y, z\} : x, y, z \in S\} = S^{(3)}$, i.e. let $\mathcal{A}$ be the set of all three-sets of points from S.

Let $f : \mathcal{A} \to \{\bullet, \blacksquare\}$ be defined in the following way: $f(\{x, y, z\}) = \bullet$ if and only if x, y, and z are not collinear. Otherwise, $f(\{x, y, z\}) = \blacksquare$.

By Ramsey's theorem, there is a monochromatic infinite subset $A \subset S$.

If $f(\{x, y, z\}) = \bullet$, for all $\{x, y, z\} \in A^{(3)}$, then no three points of A lie on a line.

Let $f(\{x, y, z\}) = \blacksquare$, for all $x, y, z \in A^{(3)}$. Let $a, b \in A$ and let L be the line through a and b. Let c be any other point in A. Since a, b, c are collinear, c lies on L. Therefore, $A \subseteq L$. □

Exercise 2.20 A function is convex if its epigraph (the set of points on or above the graph of the function) is a convex set, i.e. if the line segment that connects two points in the epigraph also belongs to the epigraph. A function f is concave if the function $-f$ is convex. Equivalently, a function f is concave if its hypograph (the set of points on or below the graph of the function) is a convex set.

Note that any linear function is both convex and concave. For the purpose of this problem we will take a linear function to be convex.

We two-colour $\mathbb{N}^{(3)}$ in the following way:

- Colour $\{i, j, k\}$, $i < j < k$, by $\blacksquare$ if the points (i, y_i), (j, y_j), and (k, y_k) induce a convex function.

- Colour $\{i,j,k\}$, $i < j < k$, by • if the points (i, y_i), (j, y_j), and (k, y_k) induce a concave function.

By Ramsey's theorem, there is an infinite monochromatic set $A = \{i_1 < i_2 < i_3 < \ldots\} \subseteq \mathbb{N}$. This means that, for any $p, q, r \in \mathbb{N}$, the set $\{i_p, i_q, i_r\}$ is always of the same colour. Say that colour is ■.

Let F be the function induced by the sequence of points (i_1, y_{i_1}), (i_2, y_{i_2}), $(i_3, y_{i_3}), \ldots$, Suppose that the epigraph of F is not convex. This means that there are points X and X' in the epigraph of F such that the line segment $\overline{XX'}$ intersect the hypograph of F, i.e. enters and exists the hypograph, possibly multiple times. Let Y and Y' be the intersection points of the graph of F and the line segment $\overline{XX'}$ such that the line segment $\overline{XY}$ is in the epigraph of F, the line segment $\overline{YY'}$ is in the hypograph of F, and there is a point Z' in $\overline{XX'}$ such that the line segment $\overline{Y'Z'}$ is in the epigraph of F.

Let $f(t) = at + b$ be an equation of the line ℓ that passes through the points X and X'. Let $j, k \in \mathbb{N}$ and $u \in (i_j, i_{j+1})$ and $u' \in (i_{k-1}, i_k)$ be such that $Y = (u, f(u))$ and $Y' = (u', f(u'))$. Observe that, since $\overline{YY'}$ is in the hypograph of F, the point(s) $(i_{j+1}, y_{i_{j+1}})$ and $(i_{k-1}, y_{i_{k-1}})$ are above the line ℓ. Since the line ℓ intersects the line segment with the end points (i_j, y_{i_j}) and $(i_{j+1}, y_{i_{j+1}})$ and the line segment with the end points $(i_{k-1}, y_{i_{k-1}})$ and (i_k, y_{i_k}), we conclude that both (i_j, y_{i_j}) and (i_k, y_{i_k}) are below the line ℓ. This means that the line ℓ intersects the line segment with the end points $(i_{j+1}, y_{i_{j+1}})$ and (i_k, y_{i_k}) at the point $Y'' = (v, f(v))$, for some $v \in (i_{j+1}, i_k)$.

Now, all points $(t, f(t))$, $t \in (i_j, u) \cup (v, i_k)$, belong to the epigraph of the function induced by the points (i_j, y_{i_j}), $(i_{j+1}, y_{i_{j+1}})$, (i_k, y_{i_k}) and all points $(t, f(t))$, $t \in (u, v)$, belong to its hypograph. This contradicts the fact that the epigraph of the function induced by the points (i_j, y_{i_j}), $(i_{j+1}, y_{i_{j+1}})$, and (i_k, y_{i_k}) is convex.

Therefore, the function F is convex.

Now we consider the case that for any p, q, and r the set $\{i_p, i_q, i_r\}$ is always coloured •. In particular, this means that for any p, q, and r the function induced by (i_p, y_{i_p}), (i_q, y_{i_q}), and (i_r, y_{i_r}) is concave. By definition, for any p, q, and r, the function induced by $(i_p, -y_{i_p})$, $(i_q, -y_{i_q})$, and $(i_r, -y_{i_r})$ is convex, and, as we have already seen, the function g induced by the sequence of points $(i_1, -y_{i_1}), (i_2, -y_{i_2}), (i_3, -y_{i_3}), \ldots$ is convex. It follows that the function $-g$ induced by the sequence of points $(i_1, y_{i_1}), (i_2, y_{i_2}), (i_3, y_{i_3}), \ldots$ is concave. □

Exercise 2.21 Let a three-colouring c of $\mathbb{N}^{(3)}$ be defined in the following way: For $i, j, k \in \mathbb{N}$, $i < j < k$,

- $c(\{i, j, k\}) = ▲$ if $y_i = y_j = y_k$.

- $c(\{i,j,k\}) = \blacksquare$ if not all of y_i, y_j, y_k are mutually equal and if the points (i, y_i), (j, y_j), and (k, y_k) induce a convex function.

- $c(\{i,j,k\}) = \bullet$ if not all of y_i, y_j, y_k are mutually equal and if the points (i, y_i), (j, y_j), and (k, y_k) induce a concave function.

By Ramsey's theorem there is an infinite c-monochromatic set $A = \{i_1 < i_2 < i_3 < \cdots\} \subseteq \mathbb{N}$.

If, for any $\{p < q < r\} \subset \mathbb{N}$, $c(\{i_p, i_q, i_r\}) = \blacktriangle$, then $y_{i_p} = y_{i_q} = y_{i_r}$ and the induced function is a constant.

If, for any $\{p < q < r\} \subset \mathbb{N}$, $c(\{i_p, i_q, i_r\}) = \blacksquare$, then y_p, y_q, y_r are not mutually equal and the points (i_p, y_{i_p}), (i_q, y_{i_q}), and (i_r, y_{i_r}) induce a convex function. By Exercise 2.20, the function induced by (i_1, y_{i_1}), (i_2, y_{i_2}), (i_3, y_{i_3}), ..., is convex.

If, for any $\{p < q < r\} \subset \mathbb{N}$, $c(\{i_p, i_q, i_r\}) = \bullet$, then y_p, y_q, y_r are not mutually equal and the points (i_p, y_{i_p}), (i_q, y_{i_q}), and (i_r, y_{i_r}) induce a concave function. By Exercise 2.20, the function induced by (i_1, y_{i_1}), (i_2, y_{i_2}), (i_3, y_{i_3}), ..., is concave. $\qquad\square$

7.2   CHAPTER 3: VAN DER WAERDEN'S THEOREM

Exercise 3.1

1. (a) Observe that if $a, a + d, a + 2d, a + 3d$ is contained in $[1, n]$, then for any $b \in [1, a)$ the arithmetic progression $b, b + d, b + 2d, b + 3d$ is also contained in $[1, n]$. Hence, the question is to find the largest $a \in [1, n]$ with the the property that $a, a + d, a + 2d, a + 3d$ is contained in $[1, n]$. Clearly, if a is the largest possible then $a + 3d = n$, what is the same as $a = n - 3d$. It follows that $s(d) = n - 3d$.

   (b) Say that $d = m = \frac{n-1}{3}$. From $1 + 3 \cdot d = 1 + 3 \cdot \frac{n-1}{3} = 1 + (n - 1) = n$, it follows that the arithmetic progression $1, 1 + d, 1 + 2d, 1 + 3d$ is contained in $[1, n]$. What about the arithmetic progression $1, 1 + (d + 1), 1 + 2(d + 1), 1 + 3(d + 1)$? From $1 + 3 \cdot (d + 1) = 1 + 3 \cdot \left(\frac{n-1}{3} + 1\right) = 1 + (n - 1) + 3 = n + 3 > n$, it follows that this arithmetic progression is not contained in $[1, n]$. Hence the maximum value of d is $m = \frac{n-1}{3}$.

   (c) It follows, from (a) and (b), that $A_n(4) = \sum_{d=1}^{\frac{n-1}{3}} s(d) = \sum_{d=1}^{\frac{n-1}{3}} (n - 3d) = \frac{(n-1)(n-2)}{6} < \frac{n^2}{2 \cdot 3}$.

2. (a) Observe that if $a, a + d, \ldots, a + (k - 1)d$ is contained in $[1, n]$, then for any $b \in [1, a)$ the arithmetic progression $b, b + d, \ldots, b + (k -$

1)d is also contained in $[1,n]$. Hence, the question is to find the largest $a \in [1,n]$ with the property that $a, a+d, \ldots, a+(k-1)d$ is contained in $[1,n]$. Clearly, if a is the largest possible then $a+(k-1)d = n$ what is the same as $a = n-(k-1)d$.

(b) Say that $d = \left\lfloor \frac{n-1}{k-1} \right\rfloor$. From $1+(k-1)\cdot d = 1+(k-1)\cdot \left\lfloor \frac{n-1}{k-1} \right\rfloor \le 1+(k-1)\cdot \frac{n-1}{k-1} = 1+(n-1) = n$ we conclude that the arithmetic progression $1, 1+d, \ldots, 1+(k-1)d$ is contained in $[1,n]$.

What about the arithmetic progression $1, 1+(d+1), \ldots, 1+(k-1)(d+1)$? From $1+(k-1)\cdot(d+1) = 1+(k-1)\cdot\left(\left\lfloor \frac{n-1}{k-1} \right\rfloor +1\right) > 1+(k-1)\cdot \frac{n-1}{k-1} = 1+(n-1) = n$, it follows that this arithmetic progression is not contained in $[1,n]$. Hence the maximum value of d is $\left\lfloor \frac{n-1}{k-1} \right\rfloor$.

(c) It follows, from (a) and (b), that $A_n(k) = \sum_{d=1}^{\left\lfloor \frac{n-1}{k-1} \right\rfloor} s(d) = \sum_{d=1}^{\left\lfloor \frac{n-1}{k-1} \right\rfloor}(n-(k-1)d) = \frac{1}{2}\cdot\left\lfloor \frac{n-1}{k-1} \right\rfloor \cdot \left(2n-(k-1)\cdot\left(\left\lfloor \frac{n-1}{k-1} \right\rfloor +1\right)\right) \le \frac{1}{2}\cdot \frac{n-1}{k-1}\cdot\left(2n-(k-1)\cdot\frac{n-1}{k-1}\right) = \frac{n^2-1}{2\cdot(k-1)} < \frac{n^2}{2\cdot(k-1)}.$ □

Exercise 3.2 Recall that in Theorem 3.4.1 and Exercise 3.1 we counted the number of arithmetic progressions of the given length in a given interval.

Let $m, n \in \mathbb{N}$, $m < n$.

Let $r \in \mathbb{N}$ be such that $(m-1)r \le n-1 < (m-1)(r+1)$, i.e. let r be the largest number for which there is an m-term arithmetic progression contained in $[0, n-1]$ with common difference r. It follows that, for any $d \in [1,r]$ and any $a \in [0, n-1-d(m-1)]$, the m-term arithmetic progression $a, a+d, \ldots, a+(m-1)d$ is contained in $[0, n-1]$. Therefore the number of m-term arithmetic progressions contained in $[0, n-1]$ is $M = \sum_{d=1}^{r}(n-(m-1)d) = nr - \frac{1}{2}(m-1)r(r+1)$.

This fact, together with the fact that $n-1 < (m-1)(r+1)$, implies that

$$M > r \cdot \frac{2(n-1)-(m-1)(r+1)}{2} > \frac{n-m}{m-1}\cdot\frac{n-1}{2} = \frac{n^2-(1+m)n+m}{2(m-1)}.$$

Observe that $\frac{n^2-(1+m)n+m}{2(m-1)} > \frac{n^2}{4(m-1)}$ is equivalent to $\frac{n^2}{2} > (1+m)n - m$, and recall that the quadratic polynomial $n \mapsto \frac{n^2}{2}$ will, at some point, overtake the linear function $n \mapsto (1+m)n - m$, regardless to the value of the natural number m.

Therefore, for all large enough natural numbers n, $M > \frac{n^2}{4(m-1)} > \frac{n^2}{4m}$. □

Exercise 3.3 We divide the word $\mathcal{W}$ in an infinite sequence of two-letter blocks and enumerate each block:

$$\mathcal{W} = \boxed{**}_{1} \boxed{**}_{2} \cdots \boxed{**}_{n} \cdots$$

Observe that each block is one of the following nine possible two–letter words:

$$\boxed{AA}\ \boxed{AB}\ \boxed{AC}\ \boxed{BA}\ \boxed{BB}\ \boxed{BC}\ \boxed{CA}\ \boxed{CB}\ \boxed{CC}$$

By associating the number of the block with the two-letter word contained in the block, we obtain a nine-colouring of the set of natural numbers.

By van der Waerden's theorem there is a monochromatic arithmetic progression of the length 1,000:

$$\mathcal{W} = \cdots \underset{a_1}{\boxed{XY}} \underbrace{\cdots}_{2d} \underset{a_2 = a_1 + d}{\boxed{XY}} \underbrace{\cdots}_{2d} \cdots \underset{a_{1000} = a_1 + 999d}{\boxed{XY}} \cdots$$

Observe that if the common difference of the monochromatic progression is d then there are d two-letter blocks between two consecutive appearances of the word XY. Hence each word that separates consecutive appearances of XY has length $2d$. □

Exercise 3.4

1. Suppose that there is only a finite number of monochromatic k-term arithmetic progressions in colour 1. Then there is some natural number N such that the interval $[1, N]$ contains all of those k-term arithmetic progressions. Divide the set $\{N + 1, N + 2, \ldots\}$ into the consecutive blocks of length $W(2; k)$ to obtain $\{N + 1, N + 2, \ldots\} = \cup_{i=0}^{\infty}[N + 1 + i \cdot W(2;k), N + (i + 1) \cdot W(2;k)]$.

   We fix $i \in \{0, 1, 2, \ldots\}$. By van der Waerden's theorem, the interval $[N + i \cdot W(2;k) + 1, N + (i + 1) \cdot W(2;k)]$ must contain a monochromatic k-term arithmetic progression. Since there are no k-term monochromatic in colour 1 in the set $\{N + 1, N + 2, \ldots\}$, this arithmetic progression must be in colour 2.

   Since this is true for any $i \in \{0, 1, 2, \ldots\}$, we conclude that there is an infinite number of k-term arithmetic progressions in colour 2.

2. Suppose that $r \geq 2$ is an integer and an r-colouring $c : \mathbb{N} \to [1, r]$ is given.

   Suppose that, for each of the colours $1, \ldots, r - 1$, there is only a finite number of monochromatic k-term arithmetic progressions in that particular colour. Then there is some natural number N such that the interval $[1, N]$ contains all of k-term arithmetic progressions that are monochromatic in one of the $1, 2, \ldots, r - 1$ colours.

Next divide the set $N + 1, N + 2, \ldots$ into the consecutive blocks of length $W(r; k)$, where $W(r; k)$ is the van der Waerden number for r colours and the k-term monochromatic arithmetic progressions:

$$\{N + 1, N + 2, \ldots\} = \cup_{i=0}^{\infty} [N + i \cdot W(r; k) + 1, N + (i + 1) \cdot W(r; k)].$$

We fix $i \in \{0, 1, 2, \ldots\}$. By van der Waerden's theorem, the interval $[N + i \cdot W(r; k) + 1, N + (i + 1) \cdot W(r; k)]$ must contain a monochromatic k-term arithmetic progression. Since there are no k-term monochromatic in any of the colours $1, \ldots, r - 1$ in the set $\{N + 1, N + 2, \ldots\}$, this arithmetic progression must be in the colour r.

Since this is true for any $i \in \{0, 1, 2, \ldots\}$, we conclude that there is an infinite number of k-term arithmetic progressions in the colour r.

3. See Example 3.5.2. □

Exercise 3.5 We start by counting how many c-monochromatic pairs there are, i.e. by determining the cardinality of the set $M_c = \{\{x, y\} : x \neq y \text{ and } c(x) = c(y)\}$.

Since, $\binom{a}{2} + \binom{b}{2} < \binom{a+b}{2}$, we conclude that a "large" colour class generates more elements of M_c than a couple of "small" colour classes. Hence, the largest possible value for the number M_c is obtained if all colour classes are as large as possible, i.e. if each colour class is of the size εn. From $\varepsilon \cdot |C| = n$ it follows that, in this case, we use $|C| = \frac{1}{\varepsilon}$ colours. Since each colour class of the size εn generates $\binom{\varepsilon n}{2}$ monochromatic pairs, we conclude that $|M_c| \leq \frac{1}{\varepsilon} \cdot \binom{\varepsilon n}{2} < \frac{\varepsilon n^2}{2}$.

Recall that we chose the integer n so that there are at least $\frac{n^2}{4m}$ m-term arithmetic progressions P contained in the interval $[0, n - 1]$. Also recall that any pair $\{x, y\} \subseteq [0, n - 1]$ can belong to at most $\frac{m(m-1)}{2}$ m-term arithmetic progressions contained in $[0, n - 1]$.

Suppose that each m-term arithmetic progressions in $[0, n - 1]$ contains an element from M_c. This assumption implies that

$$\sum_{\{x,y\} \in M_c} \#(m\text{-term arithmetic progressions that contain } \{x, y\})$$

must be greater than or equal to the number of all m-term arithmetic progressions.

It follows that $|M_c| \cdot \frac{m(m-1)}{2} \geq \frac{n^2}{4m}$, which is the same as $|M_c| \geq \frac{n^2}{2m^2(m-1)}$. But, since $\varepsilon < \frac{1}{m^2(m-1)}$, this contradicts the fact that $|M_c| < \frac{\varepsilon n^2}{2}$.

Therefore there must be an m-term c-rainbow arithmetic progression, i.e. an arithmetic progression that does not contain a monochromatic pair. □

Exercise 3.6 Let $\chi : \mathbb{N} \to \{0,1\}$ be a two-colouring of positive integers and let $k \in \mathbb{N}$. We define a new colouring $\chi' : \mathbb{N} \to \{0,1\}$ by $\chi'(i) = \chi(3i)$, $i \in \mathbb{N}$. By van der Waerden's theorem, there are $a, d \in \mathbb{N}$ such that $\chi'(a) = \chi'(a+d) = \chi'(a+2d) = \ldots = \chi'(a+(k-1)d)$. By definition of the two-colouring χ' this implies that $\chi(3a) = \chi(3a+3d) = \chi(3a+2\cdot3d) = \ldots = \chi(3a+(k-1)\cdot3d)$. Therefore the k-term monochromatic progression $3a, 3a+3d, 3a+2\cdot3d, \ldots, 3a+(k-1)\cdot3d$ is χ-monochromatic. Note that the common difference of this arithmetic progression is $3d$, a number divisible by 3. □

Exercise 3.7 We prove the claim by mathematical induction on k, the number of colours.

For $k = 1$ and any $l \in \mathbb{N}$ we take $S(1,l) = l$.

Assume that the claim is true for a fixed $k \geq 1$ and any $l \in \mathbb{N}$. We fix $l \in \mathbb{N}$ and denote by $S(k,l)$ the corresponding number. Next we set $M = W(k+1; (l-1)S(k,l)+1)$, where $W(*;*)$ denotes a van der Waerden number.

Let the set of integers $[1,M]$ be $(k+1)$-coloured. Then, by van der Waerden's theorem, there is a $((l-1)S(k,l)+1)$-term monochromatic arithmetic progression $a, a+d, \ldots, a+(l-1)S(k,l)d$.

For every $x \in [1, S(k,l)d]$, this monochromatic arithmetic progression contains the l-term arithmetic progression $a, a+xd, \ldots, a+(l-1)xd$. If for one of the numbers x, the difference xd is coloured by the same colour as the corresponding l-term arithmetic progression, we are done.

Otherwise, the $S(k,l)$-term arithmetic progression $d, 2d, \ldots, S(k,l)d$ is coloured in (at most) k colours. We apply the inductive hypothesis to conclude that the k-coloured set $\{d, 2d, \ldots, S(k,l)d\} \subset [1, M]$ contains an arithmetic progression of length l with the property that its common difference and all its terms are of the same colour. Hence we can take $M = S(k+1,l)$. □

Exercise 3.8

1. (a) Let $M > 0$ and let $i \in \mathbb{N}$ be such that $i > \log_2 M$. Then $2^{i+1} - 2^i = 2^i \cdot (2-1) = 2^i > 2^{\log_2 M} = M$. Hence, there are two consecutive powers of 2 with the gap between them greater than M. Since M was arbitrarily, it follows that the set of the powers of 2 is not syndetic.

   (b) Let $A = \{a_1, a_2, a_3, \ldots\} = \{3, 5, 6, 9, 10, 12, 15, \ldots\}$ be the set of all positive integers that are divisible by 3 or by 5. We observe that any interval of the form $(5k, 5(k+1))$ contains at least one number divisible by 3 and conclude that $1 \leq a_{i+1} - a_i \leq 3$. Therefore, this set is syndetic.

2. Suppose that $B = \{b_1 < b_2 < \cdots\}$ is not syndetic. Then, for any $M > 0$, there are two consecutive elements of B, say b_i and b_{i+1}, such that $b_{i+1} - b_i > M$. It follows that the set A, the complement of the set B in $\mathbb{N}$, contains the interval $[b_i + 1, b_{i+1} - 1]$, an interval of the length $b_{i+1} - 1 - (b_i + 1) + 1 = b_{i+1} - b_i - 1 \geq M$. Therefore, if B is not syndetic then A contains arbitrarily long intervals.

3. (a) See Observation 3.4.1.

   (b) As above, $A = \{a_1, a_2, a_3, \ldots\}$ is the set of all positive integers that are divisible by 3 or by 5. Recall that, for all $i \in \mathbb{N}$, $a_{i+1} - a_i \leq 3$.

   We partition (colour) the set of all positive integers in the following way: $C_0 = A = \{a_1, a_2, a_3, \ldots\}$, $C_1 = \{a_1 + 1, a_2 + 1, a_3 + 1, \ldots\} \backslash C_0$, and $C_2 = \{a_1 + 2, a_2 + 2, a_3 + 2, \ldots\} \backslash (C_0 \cup C_1)$. Observe that if a is a positive integer that does not belong to the set A then its remainder, after dividing by 3, must be either 1 or 2. Hence, a is in C_1 or C_2, which proves that this partition is a three colouring of $\mathbb{N}$. By van der Waerden's theorem, for any $k \in \mathbb{N}$, there is a monochromatic k-term arithmetic progression, $a, a + d, \ldots, a + (k - 1)d$. There are three possible cases:

   A. $\{a, a + d, \ldots, a + (k - 1)d\} \subset C_0 = A$
   B. If $\{a, a + d, \ldots, a + (k - 1)d\} \subset C_1$ then $(a - 1), (a - 1) + d, \ldots, (a - 1) + (k - 1)d$ is an arithmetic progression and $\{(a - 1), (a - 1) + d, \ldots, (a - 1) + (k - 1)d\} \subset A$.
   C. If $\{a, a + d, \ldots, a + (k - 1)d\} \subset C_2$ then $(a - 2), (a - 2) + d, \ldots, (a - 2) + (k - 1)d$ is an arithmetic progression and $\{(a - 2), (a - 2) + d, \ldots, (a - 2) + (k - 1)d\} \subset A$.

   Therefore for any $k \in \mathbb{N}$, the set A contains a k-term monochromatic progression. *Note*: The set A contains the k-term arithmetic progression $3, 6, 9, \ldots, 3k$.

4. Let $A = \{a_1 < a_2 < a_3 < \cdots\}$ be a syndetic set and let M be the least positive integer such that, for all $i \in \mathbb{N}$, $a_{i+1} - a_i \leq M$. We partition (colour) the set of all positive integers in the following way: $C_0 = A = \{a_1, a_2, a_3, \ldots\}$, $C_1 = \{a_1 + 1, a_2 + 1, a_3 + 1, \ldots\} \backslash C_0$, $\ldots$, $C_i = \{a_1 + i, a_2 + i, a_3 + i, \ldots\} \backslash (C_0 \cup C_1 \cup \cdots \cup C_{i-1})$, $\ldots$, $C_{M-1} = \{a_1 + M - 1, a_2 + M - 1, a_3 + M - 1, \ldots\} \backslash (C_0 \cup C_1 \cup \cdots \cup C_{M-2})$.

Observe that if a is a positive integer that does not belong to the set A then its remainder, after dividing by M, belongs to the set $\{1, 2, \ldots, M - 1\}$. Hence, a belongs to C_i for some $i \in \{1, \ldots, M - 1\}$, which proves that this partition is an M colouring of $\mathbb{N}$. By van der Waerden's theorem, for any $k \in \mathbb{N}$, there is monochromatic k-term arithmetic progression $a, a+d, \ldots, a+(k-1)d$. We distinguish two cases:

(a) $\{a, a+d, \ldots, a+(k-1)d\} \subset C_0 = A$
(b) If for some $i \in \{1, 2, \ldots, M-1\}$, $\{a, a+d, \ldots, a+(k-1)d\} \subset C_i$ then $(a-i), (a-i)+d, \ldots, (a-i)+(k-1)d$ is an arithmetic progression and $\{(a-i), (a-i)+d, \ldots, (a-i)+(k-1)d\} \subset A$.

Therefore for any $k \in \mathbb{N}$, the set A contains a k-term monochromatic progression.

5. Suppose that neither of the classes contains arbitrarily long intervals of consecutive integers. This implies that both classes are syndetic sets and contain arithmetic progressions of arbitrary length. □

Exercise 3.9 If the program runs forever, it would create two syndetic sets, A and B. By Exercise 3.8, part 4, each of those sets would contain a 10^9-term arithmetic progression. Say $\{a, a+d, a+2d, \ldots, a+(10^9-1)d\} \subset A$ and $\{b, b+\delta, b+2\delta, \ldots, b+(10^9-1)\delta\} \subset B$, where a, b, d, and δ are positive integers. Observe that both of those arithmetic progressions are finite sets, so there are integers N and M such that $\{a, a+d, a+2d, \ldots, a+(10^9-1)d\} \subseteq A(N)$ and $\{b, \delta+d, b+2\delta, \ldots, b+(10^9-1)\delta\} \subseteq B(M)$. Next we consider the set of 10^9 points $T = \{T_j = (a+jd, b+j\delta) : j \in [0, 10^9-1]\}$. Since, for any $n, m \in [0, 10^9-1]$, $n \neq m$, $\frac{(b+n\delta)-(b+m\delta)}{(a+nd)-(b+md)} = \frac{(n-m)\delta}{(n-m)d} = \frac{\delta}{d} > 0$, it follows that all points from the set T lie on a line with a positive slope. Moreover, since for any $j \in [1, 10^9-1]$, $|T_{j-1}T_j|^2 = ((a+jd)-(a+(j-1)d))^2 + ((b+j\delta)-(b+(j-1)\delta))^2 = d^2+\delta^2$, we conclude that the set T is with the required property. Therefore, the program would stop at, or before, the $\max\{N, M\}$ step. □

Exercise 3.10 Label the cells $1, 2, 3, \ldots$

1. In their first 501 turns, Maker puts a triangle into any 501 even cells to claim a sequence of cells $t_1 < t_2 < \ldots < t_{501}$. In the next turn, Maker claims an even cell t_{502} so big that the cell $\frac{t_1+t_{502}}{2}$ is bigger (further to the right) than any cell occupied by a circle. Observe that there are 1,002 empty cells that Maker may use to create a three-term arithmetic progression.

2. Maker wins by using the same strategy as above.

3. Let $N = N\left(k, \frac{1}{1,001}\right)$ be the integer guaranteed by Szemerédi's theorem, i.e. N is with the property that any set $M \subset [1, N]$ that has at least $\frac{N}{1,001}$ elements contains a k-term arithmetic progression.

Say that Maker starts first and, as long as possible, in each turn claims a cell in $[1, N]$. Observe that Breaker cannot prevent Maker to claim at least $\frac{N}{1,001}$ cells in $[1, N]$ and win.

If Breaker starts first, Maker chooses a cell a that is further away than any of 1,000 cells claimed by Breaker in their first turn. After that, in each turn Maker claims a cell in $[a + 1, a + N]$.

Note: For more about the van der Waerden and Ramsey type games see [7].

□

Exercise 3.11 Let $k \in \mathbb{N}$. Let $\varepsilon, \theta \in (-0.5, 0.5]$ be such that $k\alpha = [k\alpha] + \varepsilon$ and $(k + 1)\alpha = [(k + 1)\alpha] + \theta$. Then, from $0 \le [(k + 1)\alpha] - [k\alpha] = \alpha + (\varepsilon - \theta) \le \alpha + 1$, it follows that S_α is a syndetic set and therefore contains long arithmetic progression.

□

Exercise 3.12 Let $a, d \in \mathbb{N}$, $a < d$ and $\frac{a+0.5}{d} < 0.5$. Since $\beta = \frac{\alpha}{\alpha-1}$ is a positive irrational number, by the density of $\{k\beta\} = k\beta - \lfloor k\beta \rfloor$ in $[0, 1]$, see Exercise 2.10, there is $k \in \mathbb{N}$ such that $\frac{a}{d} < k\beta - \lfloor k\beta \rfloor < \frac{a+0.5}{d} < 0.5$. It follows that $0 < dk\beta - (a + d\lfloor k\beta \rfloor) < 0.5$, which implies that $[dk\beta] = a + d[k\beta]$. Hence S_β intersect the (arbitrary) infinite arithmetic progression $a, a + d, a + 2d, \ldots$, and S_α, the complement of S_β, cannot contain an infinite arithmetic progression.
□

Exercise 3.13 Let c be a three-colouring of the interval $[1, 18]$. Say that the first colour is red, the second colour is blue, and the third colour is green. If there are at least two elements coloured red, then there is a red two-term arithmetic progression.

Suppose that there is only one element $i \in [1, 18]$ coloured red. If $i \in [1, 9]$ then $[i + 1, 18]$ contains at least nine consecutive integers coloured blue or green. Since $W(2; 3) = 9$ there is monochromatic three-term arithmetic progression contained in $[i + 1, 18]$. If $i \in [10, 18]$ then $[1, i - 1]$ contains at least nine consecutive integers coloured blue or green. Since $W(2; 3) = 9$ there is monochromatic three-term arithmetic progression contained in $[1, i - 1]$.

If there is no element coloured red, then the interval $[1, 18]$ is two-coloured. Since $W(2; 3) = 9$ there is monochromatic three-term arithmetic progression contained in $[1, 18]$.

Therefore any three-colouring of $[1, 18]$ contains a two-term arithmetic progression in the first colour or a three-term arithmetic progression in the second or third colour.

In 1971, Tom Brown proved that $W(3; 2, 3, 3) = 14$ [13].

□

Exercise 3.14 Suppose that $k \equiv \pm 1 \pmod 6$. The claim is obviously true for $k = 1$. Hence suppose that $k \geq 5$. Let $l \geq 1$ be such that $k = 6l + 1$ or $k = 6l - 1$. Observe that k is an odd number.

To show that $W(3; k, 2, 2) = 3k$, it is sufficient to prove that $W(3; k, 2, 2) \leq 3k$ and $W(3; k, 2, 2) \geq 3k$.

To prove that $W(3; k, 2, 2) \leq 3k$ we need to show that any three-colouring (say, red, blue, and green) of the interval $[1, 3k]$ will yield a red k-term arithmetic progression or a blue two-term arithmetic progression or a green two-term arithmetic progression.

Any colouring of $[1, 3k]$ that contains two blue or two green elements will yield a blue two-term arithmetic progression or a green two-term arithmetic progression.

Hence we consider a three colouring $c : [1, 3k] \to \{$red, blue, green$\}$ that colours at most one element blue and at most one element green.

Observe that, if one of the colours blue or green is not used, there is a sequence of at least k consecutive red elements, i.e. there is a red k-term arithmetic progression.

Assume that $c : [1, 3k] \to \{$red, blue, green$\}$ is a colouring such that there is a unique $x \in [1, 3k]$ such that $c(x) =$ blue and a unique $y \in [1, 3k]$ such that $c(y) =$ green. Suppose that $y < x$.

If $y > k$ or $x < 2k$ then there is a sequence of at least k consecutive red elements between $[1, y - 1]$ or between $[x + 1, 3k]$, i.e. there is a red k-term arithmetic progression.

Suppose that $1 \leq y \leq k < 2k \leq x \leq 3k$. Observe that the number of elements between x and y is $d = x - y - 1$.

Note that if $y < k$ or $x > 2k$ then $d \geq k$, which means that there is a sequence of at least k consecutive red elements between y and x.

The only remaining case is if $y = k$ and $x = 2k$. Observe that, since $k \equiv \pm 1 \pmod 6$, neither k nor $2k$ is divisible by 3. This implies that all elements of the k-term arithmetic progression $3, 3 + 3, \ldots, 3 + 3i, \ldots, 3 + 3(k - 1) = 3k$ are coloured red.

Therefore $W(3; k, 2, 2) \leq 3k$.

To prove $W(3; k, 2, 2) \geq 3k$, we must show that there is a three-colouring of $[1, 3k - 1]$ which does not produce a red k-term arithmetic progression or a blue two-term arithmetic progression or a green two-term arithmetic progression.

Let the colouring $c : [1, 3k - 1] \to \{$red, blue, green$\}$ be defined in the following way: $c(k) =$ green, $c(2k) =$ blue, and $c(x) =$ red, otherwise. Let $a, a + d, \ldots, a + id, \ldots, a + (k - 1)d$ be a k-term arithmetic progression contained in $[1, 3k - 1]$. Observe that $d \in \{1, 2, 3\}$. Otherwise $a + (k - 1)d \geq 1 + 4(k - 1) = 4k - 3$ which is greater than $3k - 1$, for $k \geq 3$.

Since any set of k consecutive integers in $[1, 3k - 1]$ must contain the integer k or the integer $2k$ we conclude that an arithmetic progression contained in $[1, 3k - 1]$ with the common difference $d = 1$ cannot be c-monochromatic.

If the arithmetic progression $a, a + 2, \ldots, a + 2i, \ldots, a + 2(k - 1)$ is contained in $[1, 3k - 1]$ then $a + 2(k - 1) \leq 3k - 1$ implies that $a \leq k + 1$.

If a is an odd number then, for some $i \in \{0, 1, \ldots, k - 1\}$, $a + 2i = k$ and the corresponding k-term arithmetic progression is not c-monochromatic.

If a is an even number then, for some $i \in \{0, 1, \ldots, k - 1\}$, $a + 2i = 2k$ and the corresponding k-term arithmetic progression is not c-monochromatic.

If the arithmetic progression $a, a + 3, \ldots, a + 3i, \ldots, a + 3(k - 1)$ is contained in $[1, 3k - 1]$ then $a + 3(k - 1) \leq 3k - 1$ implies that $a \leq 2$.

If $a = 1$ and $k = 6l + 1$, then $1 + 3 \cdot (2l) = k$ and the corresponding k-term arithmetic progression is not c-monochromatic.

If $a = 1$ and $k = 6l - 1$, then $1 + 3 \cdot (4l - 1) = 12l - 2 = 2(6l - 1) = 2k$ and the corresponding k-term arithmetic progression is not c-monochromatic.

If $a = 2$ and $k = 6l + 1$ then $2 + 3 \cdot (4l) = 12l + 2 = 2 \cdot (6l + 1) = 2k$ and the corresponding k-term arithmetic progression is not c-monochromatic.

If $a = 2$ and $k = 6l - 1$, then $2 + 3 \cdot (2l - 1) = 6l - 1 = k$ and the corresponding k-term arithmetic progression is not c-monochromatic.

Hence, the colouring c of $[1, 3k - 1]$ does not produce a red k-term arithmetic progression or a blue two-term arithmetic progression or a green two-term arithmetic progression which proves that $W(3; k, 2, 2) \geq 3k$.

Therefore, $k = \pm 1 \pmod 6$ implies that $W(3; k, 2, 2) = 3k$. □

Exercise 3.15 A $\Rightarrow$ B. Suppose **A**. Let $\varepsilon > 0$. There is $N \in \mathbb{N}$ such that $n > N$ implies $\frac{r_k(n)}{n} < \varepsilon$, or, what is the same, $r_k(n) < \varepsilon n$. Let $n > N$ and let $A \subset [1, n]$ so that $|A| > \varepsilon n$. But this means $|A| > r_k(n)$ and A contains a k-term arithmetic progression. Hence **B**.

B $\Rightarrow$ C. Suppose **B**. Let $A \subseteq \mathbb{N}$ be such that $\overline{d}(A) = \varepsilon > 0$. This means that there is a sequence $\{n_m\}$ such that $\lim_{m \to \infty} \frac{|A \cap [1, n_m]|}{n_m} = \varepsilon$. In particular, this means that there is $M_0 \in \mathbb{N}$ such that $m > M_0$ implies that $-\frac{\varepsilon}{2} < \frac{|A \cap [1, n_m]|}{n_m} - \varepsilon < \frac{\varepsilon}{2}$. Hence, for any $m > M_0$, $\frac{\varepsilon}{2} < \frac{|A \cap [1, n_m]|}{n_m}$.

Let $k \geq 3$ and let $N = N(k, \frac{\varepsilon}{2})$ be the integer guaranteed by **B**. Let $m > M_0$ be such that $n_m > N$. Then $|A \cap [1, n_m]| > \frac{\varepsilon}{2} n_m$ and $A \cap [1, n_m]$ contains a k–term arithmetic progression. Hence **C**.

C $\Rightarrow$ A. Suppose **C** and assume that **A** is not true, i.e. assume that there is $k \geq 3$ such that $\lim_{n \to \infty} \frac{r_k(n)}{n} \neq 0$. This means that there is an increasing sequence of natural numbers $\{n_m\}$ and $\varepsilon > 0$ such that, for each $m \in \mathbb{N}$, $\frac{r_k(n_m)}{n_m} \geq \varepsilon$.

Next, we consider a subsequence $\{n_{m_i}\}$ of the sequence $\{n_m\}$ so that $n_{m_1} = n_1$ and, for $i \geq 2$, $n_{m_i} > 2(n_{m_1} + n_{m_2} + \ldots + n_{m_{i-1}})$. Let, for each $i \in \mathbb{N}$, the set A_i be the largest subset of $[1, n_{m_i}]$ that does not contain a k-term arithmetic progression. Hence, $|A_i| = r_k(n_{m_i})$.

We partition the set of natural numbers in the following way: $P_1 = [1, 2n_{m_1}]$, $P_2 = [2n_{m_1} + 1, 2(n_{m_1} + n_{m_2})]$, $\ldots$, $P_i = [2(n_{m_1} + \ldots + n_{m_{i-1}}) + 1, 2(n_{m_1} + \ldots + n_{m_i})]$, $\ldots$ Observe that $|P_i| = 2n_{m_i}$.

Let B_1 be the translate of the set A_1 by $n_{m_1} + 1$, i.e. let $B_1 = \{n_{m_1} + 1 + a : a \in A_1\}$. For $i \in \mathbb{N}\backslash\{1\}$, let B_i be the translate of the set A_i by $2(n_{m_1} + \ldots + n_{m_{i-1}}) + n_{m_i} + 1$. See below.

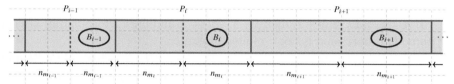

Observe that, for each $i \in \mathbb{N}$, as a translate of A_i, the set B_i does not contain a k-term arithmetic progression and that $|B_i| = r_k(n_{m_i})$. Also, observe that $\min B_{i+1} - \max B_i > n_{m_{i+1}} > 2\sum_{j=1}^{i} n_{m_j}$.

Let $B = \cup_{i=1}^{\infty} B_i$. From

$$\frac{|B \cap [1, 2(n_{m_1} + \ldots + n_{m_i})]|}{2(n_{m_1} + \ldots + n_{m_i})} = \frac{|\cup_{j=1}^{i} B_j|}{2\sum_{j=1}^{i} n_{m_j}} = \frac{\sum_{j=1}^{i} |B_j|}{2\sum_{j=1}^{i} n_{m_j}}$$

$$= \frac{\sum_{j=1}^{i} r_k(n_{m_j})}{2\sum_{j=1}^{i} n_{m_j}} \geq \frac{\sum_{j=1}^{i} \varepsilon n_{m_j}}{2\sum_{j=1}^{i} n_{m_j}} = \frac{\varepsilon}{2} > 0$$

it follows that the set B has a positive upper density. By **C**, the set B contains a k-term arithmetic progression $a, a+d, \ldots, a+(k-1)d$.

Let $i \in \mathbb{N}$ be such that $a \in B_i$.

Suppose that $a + d \in B_i$. Since B_i does not contain a k-term arithmetic progression there is $j < k - 1$ such that $a + jd \in B_i$ and $a + (j+1)d \notin B_i$. But since $0 < d < n_{m_i}$ and since $\min B_{i+1} - \max B_i > n_{m_{i+1}}$, it follows that $a + (j+1)d < \min B_{i+1}$. This contradicts the assumption that $a + (j+1)d \in B$.

So let $j > i$ be such that $a + d \in B_j$. Then $d > n_{m_j}$, since the progression "jumps" over the first half of the interval P_j, which implies that $a + 2d \notin B_j$. On the other hand, since $a, a+d \subseteq [1, 2(n_{m_1} + \ldots + n_{m_j})]$, we have that $d < 2(n_{m_1} + \ldots n_{m_j}) < n_{m_{j+1}}$. Thus $a + 2d < \min B_{j+1}$. Contradiction. Hence, **A** must hold. $\qquad\square$

Exercise 3.16 Let $A \subseteq \mathbb{N}$ be a syndetic set. Suppose that $A = \{1 = a_1 < a_2 < \cdots\}$ and that $d > 0$ is such that $0 < a_{i+1} - a_i \leq d$, for all $i \in \mathbb{N}$. This implies that each interval $[1 + (i-1)d, id]$, $i \in [1, k]$, contains at least one element ▲ of the set A:

$$ ▲ \cdots ▲ \cdots \boxed{d} \cdots ▲ \cdots \boxed{2d} \cdots ▲ \cdots ▲ \cdots \boxed{(i-1)d} \cdots ▲▲ \cdots \boxed{id} \cdots . $$

Observe that, by the definition of the upper density, $\overline{\delta}(A) = \limsup_{n\to\infty} \frac{|A \cap [1,n]|}{n} \geq \limsup_{k\to\infty} \frac{|A \cap [1,kd]|}{kd} \geq \lim_{k\to\infty} \frac{k}{kd} = \frac{1}{d}$. By Szemerédi's theorem, the syndetic set A contains arithmetic progressions of any finite length.

Note: The assumption $a_1 = 1$ was there only to simplify our calculations. Adding or taking away a single element does not change set's upper density.

□

7.3 CHAPTER 4: SCHUR'S THEOREM AND RADO'S THEOREM

Exercise 4.1 Let $a = \min A$. Then $A' = \{x - a : x \in A \backslash \{a\}\} \subset [1, n]$ and $|A'| = |A| - 1$. From $n \geq |A \cup A'| = |A| + |A'| - |A \cap A'| = 2|A| - 1 - |A \cap A'| > n - |A \cap A'|$ it follows that $A \cap A' \neq \emptyset$. Hence there are $x, y \in A$ such that $y = x - a$. □

Exercise 4.2 1. Our strategy is to try to build a red/blue colouring that avoids red Schur triples and blue three-term arithmetic progressions.

Consider the positive numbers from 1 to 7.

Colour the number 1 red. Then, to avoid a red Schur triple, 2 must be blue.

Suppose that 3 is red. Then 6 must be blue. If 4 is red, there is a red Schur triple $(1, 3, 4)$. If 4 is blue there is blue arithmetic progression $2, 4, 6$.

Suppose that 3 is blue. Then 4 must be red, 5 must be blue, 7 must be red, and 6 must be blue.

Hence we have established one colouring with the required property.

Colour 1 blue and 2 red. Then 4 must be blue, 7 must be red, 5 must be blue. None of 3 and 6 can be blue, but if both of them are red, then there is a red Schur triple $(3, 3, 6)$.

Colour 1 and 2 blue. Then 3 must be red, 6 must be blue, 4 must be red, 7 must be blue, and 5 must be red.

It follows that there are only two red/blue colourings of $[1,7]$ that avoid red Schur triples and blue three-term arithmetic progressions: R–B–B–R–B–B–R and B–B–R–R–R–B–B

2. Consider the R–B–B–R–B–B–R colouring of $[1,7]$. If we colour 8 red then there is a red Schur triple $(1,7,8)$. If we colour 8 blue then there is a blue three-term arithmetic progression $3,5,8$.

Consider the B–B–R–R–R–B–B colouring of $[1,7]$. If we colour 8 red then there is a red Schur triple $(3,5,8)$. If we colour 8 blue then there is a blue three-term arithmetic progression $6,7,8$.

Therefore any blue/red colouring of $[1,8]$ contains a red Schur triple or a blue three-term arithmetic progression. □

Exercise 4.3 Observe that there is a red/blue colouring of $[1,9]$ with no red solution of $x + y = z$ (of distinct integers) or a blue solution of $x + y = 2z$.

Next we try to build a red/blue colouring of $[1,10]$ that avoids a red (R) solution of the equation $x + y = z$ (of distinct integers) or a blue (B) solution of the equation $x + y = 2z$.

We start by observing that such a colouring has to avoid monochromatic triples $(1,2,3)$, $(2,4,6)$, and $(3,6,9)$, because these triples are both three-term arithmetic progressions and strict Schur triples.

In our attempt to avoid a red solution of $x + y = z$ (of distinct integers) or a blue solution of $x + y = 2z$, we consider all possible colourings of the triple $(2,4,6)$ that use both colours, R and B:

Case 1: If $2 = R, 4 = B, 6 = B$, then $5 = R$, because of the arithmetic progression $(4,5,6)$. This implies $7 = B$, because of $2 + 5 = 7$, as well as $8 = R$, because of the arithmetic progression $(6,7,8)$, and $10 = R$, because of the arithmetic progression $(4,7,10)$. But now, $(2,8,10)$ is a red Schur triple.

Case 2: If $2 = B, 4 = R, 6 = B$, then $10 = R$, because of the arithmetic progression $(2,6,10)$.

Now, if $3 = B$ then $1 = R$, because of the arithmetic progression $(1,2,3)$, and $9 = B$, because of $1 + 9 = 10$. But then $(3,6,9)$ is a blue arithmetic progression.

If $3 = R$ then $7 = B$, because of $3 + 4 = 7$, $5 = R$, because of the arithmetic progression $(5,6,7)$, and $8 = R$, because of the arithmetic progression $(6,7,8)$. But then $(3,5,8)$ is a red Schur triple.

Case 3: If $2 = B, 4 = B, 6 = R$ then $3 = R$, because of the arithmetic progression $(2, 3, 4)$. It follows that $9 = B$, because of $3 + 6 = 9$.

Now, if $1 = R$ then $5 = B$, because of $1 + 5 = 6$, and $7 = B$, because of $1 + 6 = 7$. But then $(5, 7, 9)$ is a blue arithmetic progression.

If $1 = B$ then $5 = R$, because of the arithmetic progression $1, 5, 9$, and $7 = B$, because of $1 + 6 = 7$. Also, $8 = B$, because of $3 + 5 = 8$. This implies $7 = R$, because of the arithmetic progression $(7, 8, 9)$, and $10 = R$. because of the arithmetic progression $(8, 9, 10)$. Now, $(3, 7, 10)$ is a red Schur triple.

Case 4: Let $2 = R, 4 = R, 6 = B$.

Now, if $3 = R$ then $5 = B$, because of $2 + 3 = 5$, and $7 = B$, because of $3 + 4 = 7$. But then $(5, 6, 7)$ is a blue arithmetic progression.

If $3 = B$ then $9 = R$, because of the arithmetic progression $3, 6, 9$, $5 = B$, because of $4 + 5 = 9$, and $7 = B$, because of $2 + 7 = 9$. Now, $(5, 6, 7)$ is a blue arithmetic progression.

Case 5: If $2 = R, 4 = B, 6 = R$ then $8 = B$, because of $2 + 6 = 8$.

Now, if $5 = R$ then $1 = B$, because of $1 + 5 = 6$, $3 = B$, because of $2 + 3 = 5$, and $7 = B$, because of $2 + 5 = 7$. But then $(1, 4, 7)$ is a blue arithmetic progression.

If $5 = B$ then $3 = R$, because of the arithmetic progression $(3, 4, 5)$, $1 = B$, because of $1 + 2 = 3$, and $7 = R$, because of the arithmetic progression $(1, 4, 7)$. Now, if $9 = B$, because of $2 + 7 = 9$), then $(1, 5, 9)$ is a blue arithmetic progression.

Case 6: If $2 = B, 4 = R, 6 = R$ then $10 = B$, because of $4 + 6 = 10$.

Now, if $3 = R$ then $1 = B$, because of $1 + 3 = 4$, $7 = B$, because of $3 + 4 = 7$, and $9 = B$, because of $3 + 6 = 9$. It follows that $8 = R$, because of the arithmetic progression $(8, 9, 10)$ and $5 = B$, because of $3 + 5 = 8$. But then $(1, 5, 9)$ is a blue arithmetic progression.

If $3 = B$ then $1 = R$, because of the arithmetic progression $1, 2, 3$, $5 = B$, because of $1 + 4 = 5$, and $7 = B$, because of $1 + 6 = 7$. But then $(3, 5, 7)$ is a blue arithmetic progression.

Therefore, it is impossible to colour $[1, 10]$ red and blue and to avoid a red Schur triple or a blue three–term arithmetic progression. □

Exercise 4.4 Let c be a finite colouring of positive integers. Define a finite colouring c' of positive integers by $c'(i) = c(2^i)$, $i \in \mathbb{N}$. By Schur's theorem, there is a c'-monochromatic triple (x_1, x_2, x_3) such that $x_1 + x_2 = x_3$.

Let $y_1 = 2^{x_1}$, $y_2 = 2^{x_2}$, and $y_3 = 2^{x_3}$. Then $y_1 y_2 = 2^{x_1} \cdot 2^{x_2} = 2^{x_1 + x_2} = 2^{x_3} = y_3$ and, since $c'(x_1) = c'(x_2) = c'(x_3)$, we have that $c(y_1) = c(y_2) = c(y_3)$. □

Exercise 4.5 Let c be a finite colouring of positive integers. Define a finite colouring c' of positive integers by $c'(i) = c(2^i - 1)$, $i \in \mathbb{N}$. By Schur's theorem, there is a c'-monochromatic triple (x_1, x_2, x_3) such that $x_1 + x_2 = x_3$.

Let $y_1 = 2^{x_1} - 1$, $y_2 = 2^{x_2} - 1$, and $y_3 = 2^{x_3} - 1$. Then $y_1 y_2 + y_1 + y_2 = (2^{x_1} - 1) \cdot (2^{x_2} - 1) + 2^{x_1} - 1 + 2^{x_2} - 1 = 2^{x_1 + x_2} - 1 = 2^{x_3} - 1 = y_3$ and, since $c'(x_1) = c'(x_2) = c'(x_3)$, we have that $c(y_1) = c(y_2) = c(y_3)$. □

Exercise 4.6

1. Let a colouring $\xi : \mathbb{N} \to [1, r]$ be given. By Schur's theorem there are $a, b, c \in \mathbb{N}$ such that $a + b - c = 0$ and $\xi(a) = \xi(b) = \xi(c)$. Hence, by definition, the point $(a, b, c) \in P$ is not marked by X.

2. The question is if, for a given finite colouring of positive integers, there is an infinite number of monochromatic Schur triples.

   Suppose that there is a colouring $\xi : \mathbb{N} \to [1, r]$ such that the set S of all ξ-monochromatic Schur triples is finite. Let p be a prime that is greater than any integer that appears in any of the ξ-monochromatic Schur triples in S. We define a colouring $\xi_p : \mathbb{N} \to [1, r]$ by $\xi_p(n) = \xi(pn)$.

   By Schur's theorem there is a ξ_p-monochromatic Schur triple, i.e. there are $a, b, c \in \mathbb{N}$ such that $a + b = c$ and $\xi_p(a) = \xi_p(b) = \xi_p(c)$. This implies that (pa, pb, pc) is a ξ-monochromatic Schur triple: $pa + pb = pc$ and $\xi(pa) = \xi(pb) = \xi(pc)$. Clearly, $(pa, pb, pc) \notin S$. This contradicts our assumption that S, the set of all ξ-monochromatic Schur triples, is finite. Therefore, the set P will contain an infinite number of coloured points. □

Exercise 4.7

1. Let $(x, y, x + y) \in S_\bullet$ and let $x \in [1, k]$.

   Observe that if $x \in [1, k]$ then we have to consider two cases: $y, x + y \in [1, 4k - 1]$ and $y, x + y \in [10k, 11k]$. (The case $y \in [1, 4k - 1]$ and $x + y \in [10k, 11k]$ is not possible. Why?)

   Case 1: Suppose that $y, x + y \in [1, 4k - 1]$.

   From $1 \le x \le k$, $x \le y$, and $x + y \le 4k - 1$ it follows that $x \le y \le 4k - x - 1$. Observe that the number of such y's is equal to $4k - x - 1 - x + 1 = 4k - 2x$. Hence, for each $x \in [1, k]$ there are $4k - 2x$ Schur triples $(x, y, x + y)$ such that $x + y \le 4k - 1$. Each of those triples is coloured $\bullet$.

   It follows that $|\{(x, y, x + y) \in S_\bullet : x \in [1, k], x \le y < x + y \le 4k - 1\}| = \sum_{x=1}^{k}(4k - 2x) = 4k^2 - 2 \cdot \frac{k(k+1)}{2} = 3k^2 - k$.

Case 2: Suppose that $y, x + y \in [10k, 11k]$.

From $1 \le x \le k$, $10k \le y$, and $x + y \le 11k$ it follows that $10k \le y \le 11k - x$. Observe that the number of such y's is equal to $11k - x - 10k + 1 = k + 1 - x$.

It follows that $|\{(x, y, x + y) \in S_\bullet : x \in [1, k], 10k \le y < x + y \le 11k\}| = \sum_{x=1}^{k}(k + 1 - x) = (k + 1) \cdot k - \frac{k(k+1)}{2} = \frac{k^2}{2} + \frac{k}{2}$.

Therefore, $|\{(x, y, x + y) \in S_\bullet : x \in [1, k]\}| = (3k^2 - k) + \left(\frac{k^2}{2} + \frac{k}{2}\right) = \frac{7k^2 - k}{2}$.

2. Let $(x, y, x + y) \in S_\bullet$ with $x \in [k + 1, 2k - 1]$.

   Observe that if $x \in [k + 1, 2k - 1]$ then $y, x + y \in [1, 4k - 1]$. (The case $10k \le y < x + y \le 11k$ is not possible. Why?)

   From $k + 1 \le x \le 2k - 1$, $x \le y$, and $x + y \le 4k - 1$ it follows that $x \le y \le 4k - x - 1$. Observe that the number of such y's is equal to $4k - x - 1 - x + 1 = 4k - 2x$. It follows that $|\{(x, y, x + y) \in S_\bullet : x \in [k + 1, 2k - 1], x \le y < x + y \le 4k - 1\}| = \sum_{x=k+1}^{2k-1}(4k - 2x)$.

   From $\sum_{x=k+1}^{2k-1} 4k = 4k(k - 1)$ and $\sum_{x=k+1}^{2k-1}(-2x) = -2\left(\sum_{x=1}^{2k-1} x - \sum_{x=1}^{k} x\right) = -3k(k - 1)$, it follows that $|\{(x, y, x + y) \in S_\bullet : x \in [k + 1, 2k - 1], x \le y < x + y \le 4k - 1\}| = 4k(k - 1) - 3k(k - 1) = k^2 - k$.

3. Observe that there is no $\bullet$-coloured Schur triple such that $x \ge 2k$. This is because $2k \le x \le 4k - 1$ and $x \le y \le 4k - 1$ imply that $c(x + y) = \blacksquare$, that $2k \le x \le 4k - 1$ and $10k \le y \le 11k$ imply that $x + y > N$, and that $10k \le x \le y \le 11k$ implies that $x + y > N$. Hence, $|S_\bullet| = \frac{7k^2 - k}{2} + k^2 - k = \frac{9k^2}{2} - \frac{3k}{2}$.

4. Let $(x, y, x + y) \in S_\blacksquare$. Observe that this implies that $x \in [4k, 5k - 1]$. It follows that, if $(x, y, x + y) \in S_\blacksquare$, then $x \in [4k, 5k - 1]$ and $x \le y < x + y \le 10k - 1$.

   From $4k \le x \le 5k - 1$, $x \le y$, and $x + y \le 10k - 1$ it follows that $x \le y \le 10k - x - 1$. Observe that the number of such y's is equal to $10k - x - 1 - x + 1 = 10k - 2x$. Hence, for each $x \in [4k, 5k - 1]$, there are $10k - 2x$ $\blacksquare$-coloured Schur triples. It follows that $|S_\blacksquare| = \sum_{x=4k}^{5k-1}(10k - 2x)$.

   From $\sum_{x=4k}^{5k-1} 10k = 10k \cdot (5k - 1 - 4k + 1) = 10k^2$ and $\sum_{x=4k}^{5k-1}(-2x) = -2\left(\sum_{x=1}^{5k-1} x - \sum_{x=1}^{4k-1} x\right) = -(9k^2 - k)$ it follows that $|S_\blacksquare| = 10k^2 - (9k^2 - k) = k^2 + k$.

5. Observe that the number of all ξ-monochromatic Schur triples is given by

$$|S_\xi| = |S_\bullet \cup S_\blacksquare| = \frac{9k^2}{2} - \frac{3k}{2} + k^2 + k = \frac{11k^2}{2} - \frac{k}{2} = \frac{N^2}{22} - \frac{N}{22}.$$

6. Recall that, by definition, $S(N) = \inf_{c:[1,N] \to \{\bullet, \blacksquare\}} \{|S_c|\}$. Since ξ is a two-colouring of $[1, N]$, it follows that $S(N) \le |S_\xi| = \frac{N^2}{22} - \frac{N}{22} < \frac{N^2}{22}$.

Note: The colouring ξ and the fact that, for $N \in \mathbb{N}$, $S(N) = \frac{N^2}{22} + O(N)$ was established by Robertson and Doron Zeilberger, an Israeli mathematician, in 1998. See [97]. □

Exercise 4.8

1. Consider the sum $x_1 + x_2 + x_3$, with $x_1, x_2, x_3 \in B$. For this sum to be less than or equal to 13, we have to have $\{x_1, x_2, x_3\} \subseteq \{1, 2\}$. But in that case $3 \le x_1 + x_2 + x_3 \le 6$, which implies that $\mathcal{L}(4)$ has no blue solution. If $\{x_1, x_2, x_3, x_4\} \subseteq R$, then $x_1 + x_2 + x_3 + x_4 \ge 3 + 3 + 3 + 3 = 12$, which implies that there is no red solution of $\mathcal{L}(5)$.

2. We try to build two sets, B, the set of all elements of $[1, 14]$ coloured blue, and R, the set of all elements of $[1, 14]$ coloured red, so that $B \cup R = [1, 14]$.

   Suppose that $1 \in B$. To avoid a blue solution to $\mathcal{L}(4)$, we must have $3 \in R$. To avoid a red solution to $\mathcal{L}(5)$, we must have $4 \cdot 3 = 12 \in B$. To avoid a blue solution to $\mathcal{L}(4)$, we must have $1 + 1 + 12 = 14 \in R$. To avoid a red solution $3 + 3 + 3 + 5 = 14$, we must have $5 \in B$. To avoid a blue solution $1 + 2 + 2 = 5$, we must have $2 \in R$. To avoid a red solution $2 + 2 + 3 + 7 = 14$, we must have $7 \in B$. Hence, so far, $\{1, 5, 7, 12\} \subseteq B$ and $\{2, 3, 14\} \subseteq R$.

   But now, we have a blue solution $1 + 1 + 5 = 7$.

   Suppose that $1 \in R$. To avoid a red solution to $\mathcal{L}(5)$, we must have $4 \in B$. To avoid a blue solution to $\mathcal{L}(4)$, we must have $3 \cdot 4 = 12 \in R$. To avoid a red solution $1 + 1 + 1 + 9 = 12$, we must have $9 \in B$. To avoid a blue solution $3 + 3 + 3 = 9$, we must have $3 \in R$. Hence, so far, $\{4, 9\} \subseteq B$ and $\{1, 3, 12\} \subseteq R$.

   But now, we have a red solution $3 + 3 + 3 + 3 = 12$.

3. In part 2, we proved that $s(4, 5) \ge 14$. In part 1, we proved that $s(4, 5) \le 14$. Hence, $s(4, 5) = 14$. □

Exercise 4.9 Say that we are given an equinumerous 3–colouring $c : [1, 9] \rightarrow$ $\{\bullet, \blacksquare, \blacktriangle\}$. Let $c(1) = \bullet$.

Case 1. If there is $n \in [2, 8]$ such that $c(n) = \blacksquare$ and $c(n+1) = \blacktriangle$ (or vice versa) then $x = 1, y = n, z = n+1$ is a rainbow solution of Schur's equation.

Case 2. Let $c(9) = \bullet$. Then, to keep the numbers coloured by $\blacksquare$ and the numbers coloured by $\blacktriangle$ non–consecutive, we must have $c(5) = \bullet$. This would imply that the numbers $2, 3, 4$ are monochromatic in one colour and the numbers $6, 7, 8$ are monochromatic in the other colour. But if we take, for example, $x = 2$, $y = 5$, and $z = 7$, we obtain a rainbow solution of Schur's equation.

In the remaining cases we take $c(9) = \blacksquare$. Observe that, under this assumption, to avoid Case 1, 8 cannot be $\blacktriangle$.

Case 3. If $c(5) = \blacktriangle$, then, to avoid Case 1 and to avoid a rainbow triple $(4, 5, 9)$, $c(4) = \blacktriangle$. To avoid Case 1, neither of 3 and 6 can be $\blacksquare$. To avoid a rainbow triple $(3, 6, 9)$, 3 and 6 must be of the same colour. That colour cannot be $\blacktriangle$, so $c(3) = c(6) = \bullet$. To avoid Case 1, we have $c(2) = \blacktriangle$ and $c(7) = c(8) = \blacksquare$. But then $(2, 5, 7)$ is a rainbow Schur triple.

Case 4. If both 5 and 8 are coloured $\blacksquare$, then one of the numbers 4, 6, and 7 must be $\blacktriangle$ and we get consecutive $\blacksquare$ and $\blacktriangle$, what was discussed in Case 1.

Case 5. If both 5 and 8 are coloured $\bullet$, then, to avoid Case 1, the numbers 6, 7 must be $\blacksquare$ and we get that $2, 3, 4$ are all $\blacktriangle$. But then $(2, 5, 7)$ is a rainbow Schur triple.

Case 6. If $c(5) = \bullet$ and $c(8) = \blacksquare$ then, to avoid rainbow $1 + 7 = 8$, $3 + 5 = 8$, and $4 + 5 = 9$, we need to colour $3, 4, 7$ by $\bullet$ or $\blacksquare$. This would leave only 2 and 6 left to be coloured by $\blacktriangle$. Impossible.

Case 7. If $c(5) = \blacksquare$ and $c(8) = \bullet$ then, to avoid rainbow $1 + 4 = 5$, $1 + 5 = 6$, and $3 + 5 = 8$, we need to colour $3, 4, 6$ by $\bullet$ or $\blacksquare$. This would leave only 2 and 7 left to be coloured by $\blacktriangle$. Impossible. □

Exercise 4.10

1. Notice that the given equation is equivalent to the equation $4x_1 - 3x_2 + 24x_3 - x_4 = 0$. Since $a_1 + a_2 + a_4 = 4 - 3 - 1 = 0$, by Rado's theorem, this equation is partition regular over $\mathbb{N}$, i.e. any finite colouring of positive integers contains a monochromatic solution of this equation. But that implies that any finite colouring of positive integers contains a monochromatic solution of the original equation. Therefore, the given equation is partition regular over $\mathbb{N}$.

2. Notice that the given equation is equivalent to the equation $2x_1 + 3x_2 - x_3 + 2x_4 - 4x_5 = 0$. Since $a_1 + a_2 + a_3 + a_5 = 2 + 3 - 1 - 4 = 0$, by Rado's theorem, this equation is partition regular over $\mathbb{N}$. It follows that the given equation is partition regular over $\mathbb{N}$. *Note:* Observe that $a_1 + a_4 +$

$a_5 = 2 + 2 - 4 = 0$. Recall that, to apply Rado's theorem, it is enough to find *one* subset of coefficients that sums to 0.

3. Yes, the given equation is partition regular over $\mathbb{N}$. Notice that the sum of the first two coefficients is 0.

4. Here $a_1 = 2$, $a_2 = -1$, $a_3 = 3$, and $a_4 = -6$. From $a_1 + a_2 = 1$, $a_1 + a_3 = 5$, $a_1 + a_4 = -4$, $a_2 + a_3 = 2$, $a_2 + a_4 = -7$, $a_3 + a_4 = -3$, $a_1 + a_2 + a_3 = 4$, $a_1 + a_2 + a_4 = -5$, $a_1 + a_3 + a_4 = -1$, $a_2 + a_3 + a_4 = -4$, and $a_1 + a_2 + a_3 + a_4 = -2$, we conclude, via Rado's theorem, that the given equation is not partition regular over $\mathbb{N}$. □

Exercise 4.11 Here $a_1 = a_2 = 1$ and $a_3 = -4$. From $a_1 + a_2 = 2$, $a_1 + a_3 = -3$, $a_2 + a_3 = -3$, and $a_1 + a_2 + a_3 = -2$ we conclude, via Rado's theorem, that the given equation is not partition regular over $\mathbb{N}$. Therefore, there is a finite colouring of positive integers that does not contain a monochromatic solution of the given equation. By the proof of Proposition 4.4.1, the following colouring would do:

Note that $|a_1| + |a_2| + |a_3| = 1 + 1 + 4 = 6$. We take $p = 7$, a prime number greater than $|a_1| + |a_2| + |a_3|$, and define a six-colouring $c : \mathbb{N} \to [1,6]$ in the following way: For $n \in \mathbb{N}$ we find $k, l \in \{0, 1, 2, \ldots\}$ and $i \in [1,6]$ such that $n = 7^k(7 \cdot l + i)$. Then, by definition, $c(n) = c(7^k(7l + i)) = i$.

If there is a c-monochromatic solution of the given equation, then $x_1 = 7^k(7l + i)$, $x_2 = 7^m(7n + i)$, and $x_3 = 7^r(7s + i)$, for some $k, l, m, n, r, s \in \{0, 1, 2 \ldots\}$ and $i \in [1,6]$. Hence, $0 = 7^k(7l + i) + 7^m(7n + i) - 4 \cdot 7^r(7s + i)$ or, what is the same $(7^k + 7^m - 4 \cdot 7^r)i + 7(7^k l + 7^m n - 4 \cdot 7^r s) = 0$. Since $i \in [1,6]$, the first term of this expression is divisible by $7^{\min\{k,m,r\}}$, but not by $7^{1+\min\{k,m,r\}}$. Hence, there is no c-monochromatic solution of the given equation. □

Exercise 4.12

1. $F_3(1,2,5) = \{1, 1+2, 1+5, 1+2+5, 2, 2+5, 5\} = \{1, 2, 3, 5, 6, 7, 8\}$.

2. Observe that Folkman's theorem guarantees the existence of a monochromatic $F_2(a,b) = \{a, b, a+b\}$, in any r-colouring of the set of natural numbers.

3. *Base Case:* We take $n(r,1) = 1$. Observe that $I = \{1\}$ is the only non-empty subset of the set $[1] = \{1\}$ and $a(I) = F_1(1) = \{1\}$.

Inductive Step:

   (a) Apply van der Waerden's theorem.

(b) Observe that $dn(r,k) \le W(r;n(r,k)+1) = \frac{N}{2}$. Define an r-colouring ξ' of $[1, n(r,k)]$ by $\xi'(j) = \xi(jd)$, for any $j \in [1, n(r,k)]$. We apply the inductive hypothesis to find $b_1, b_2, \ldots, b_k \in [1, n(r,k)]$ such that, for any non-empty subset I of the set $[1,k]$, $b(I) = \sum_{i \in I} b_i \in [1, n(r,k)]]$ and $\xi'(b(I)) = \xi'(b_{\max(I)})$. This implies that, for $a_i = b_i d$, $i \in [1, k]$, we have, for any non-empty subset I of the set $[1, k]$, $a(I) = \sum_{i \in I} a_i \in \{d, 2d, \ldots, n(r,k) \cdot d\}$ and $\xi(a(I)) = \xi(a_{\max(I)})$.

(c) Consider the sequence $a_1, a_2, \ldots, a_k, a_{k+1}$, where $a_i, i \in [1, k+1]$, are established in (a) and (b). Let I be a non-empty subset of the set $[1, k+1]$. If $\max(I) < k+1$ then, by (b), $a(I) = \sum_{i \in I} a_i \in [1, N]$ and $\xi(a(I)) = \xi(a_{\max(I)})$. If $\max(I) = k+1$ then, by (b), $\sum_{i \in I \setminus \{k+1\}} a_i \in \{d, 2d, \ldots, n(r,k) \cdot d\}$, which implies that $a(I) = \sum_{i \in I} a_i = a_{k+1} + \sum_{i \in I \setminus \{k+1\}} a_i$ is a term in the ξ-monochromatic arithmetic progression that we established in (a). Hence, $\xi(a(I)) = \xi(a_{k+1})$, which establishes that $N = n(r, k+1)$ and completes the proof of the inductive step.

4. (a) This follows from (3) and our choice of M.

(b) Yes, the colouring η is well-defined. By our choice of $a_1, a_2, \ldots, a_{r(k-1)+1}$, see Part 4a of the statement of this exercise, if $I, J \subseteq [1, r(k-1)+1]$ and $\max(I) = \max(J) = j$, then $\chi(a(I)) = \chi(a(J)) = \chi(a_j)$.

(c) Since η is an r-colouring of $[1, r(k-1)+1]$, by the pigeonhole principle, at least one of the r colour classes (pigeonholes) must contain at least k elements (pigeons). Hence, there is an η-monochromatic set $S = \{s_1, s_2, \ldots, s_k\} \subset [1, r(k-1)+1]$, with $s_i < s_j$, for $i < j$.

(d) For $i \in [1, k]$, let $a_i' = a_{s_i}$. Let $I, J \subseteq [1, k]$ be two non-empty subsets with $p = \max(I)$ and $q = \max(J)$. Then $\chi\left(\sum_{i \in I} a_i'\right) = \chi\left(\sum_{i \in I} a_{s_i}\right) = \chi(a_{s_p})$ and $\chi\left(\sum_{i \in J} a_i'\right) = \chi\left(\sum_{i \in J} a_{s_i}\right) = \chi(a_{s_q})$. But since $s_p, s_q \in S$ it follows that $\eta(s_p) = \eta(s_q)$ which is the same as $\chi(a_{s_p}) = \chi(a_{s_q})$. Therefore for any $\emptyset \ne I, J \subseteq [1, k]$ we have that $\chi\left(\sum_{i \in I} a_i'\right) = \chi\left(\sum_{i \in J} a_i'\right)$. □

Exercise 4.13

1. By definition, $Q(1; 2, 3, 4) = \{1 + \varepsilon_1 \cdot 2 + \varepsilon_2 \cdot 3 + \varepsilon_3 \cdot 4 : \varepsilon_1, \varepsilon_2, \varepsilon_3 \in \{0, 1\}\} = \{1, 1+2, 1+3, 1+4, 1+2+3, 1+2+4, 1+3+4, 1+2+3+4\} = \{1, 3, 4, 5, 6, 7, 10\}$.

2. By the pigeonhole principle, there are $a, b \in [k+1, k+(r+1)]$, $a < b$, such that $\chi(a) = \chi(b)$. Hence the one-cube $Q(a; b-a) = \{a, b\}$ is χ-monochromatic.

3. Since $a_1, i \in [1, r]$, there are r^2 possible different (a_1, i) types of χ-monochromatic one-cubes in $[k+1, k+(r+1)]$.

4. (a) Since there are r^2 possible types of χ-monochromatic one-cubes in any interval of length $r+1$, and since $[1, (r^2+1)(r+1)]$ is the union of $r^2 + 1$ such intervals, by the pigeonhole principle, there are $p, q, 0 \le p < q \le r^2$, such that the intervals $[p(r+1)+1, (p+1)(r+1)]$ and $[q(r+1)+1, (q+1)(r+1)]$ contain χ-monochromatic one-cubes of the same type.

   (b) Let p and q, $0 \le p \le q \le r^2$, be such that the intervals $[p(r+1)+1, (p+1)(r+1)]$ and $[q(r+1)+1, (q+1)(r+1)]$ contain χ-monochromatic one-cubes of the same type (a_1, i). Hence there are monochromatic one-cubes, $Q(a; a_1) \subseteq [p(r+1)+1, (p+1)(r+1)]$ and $Q(b; a_1) \subseteq [q(r+1)+1, (q+1)(r+1)]$, coloured by the same colour i.

   We consider the two-cube $Q(a; a_1, b-a)$. Observe that $Q(a; a_1, b-a) = \{a, a+a_1, a+(b-a) = b, a+a_1+(b-a) = b+a_1\} = Q(a; a_1) \cup Q(b; a_1)$ implies that the two-cube $Q(a; a_1, b-a)$ is χ-monochromatic.

5. Recall that $a_1, i \in [1, r]$. Note that $(k+1)+1+a_2 \le k+(r^2+1)(r+1)$ implies that $1 \le a_2 \le (r^2+1)(r+1)-2$. Hence the number of possible types (a_1, a_2, i) is $r \cdot ((r^2+1)(r+1)-2) \cdot r$, which is less than $(r+1)^5$.

6. Observe that the interval $[1, (r^2+1)(r+1)^6] = [1, (r+1)^5 \cdot (r^2+1)(r+1)]$ contains $(r+1)^5$ consecutive intervals of length $(r^2+1)(r+1)$. By Part 5 and the pigeonhole principle, there are $p, q \in [0, (r+1)^2 - 1]$, $p < q$, such that the intervals $[p+1, p+(r^2+1)(r+1)]$ and $[q+1, q+(r^2+1)(r+1)]$ contain χ-monochromatic two-cubes of the same type, say $Q(a; a_1, a_2) \subseteq [p+1, p+(r^2+1)(r+1)]$ and $Q(b, a_1, a_2) \subseteq [q+1, q+(r^2+1)(r+1)]$.

   We consider the three-cube $Q(a, a_1, a_2, b-a)$. Observe that $Q(a; a_1, a_2, b-a) = \{a, a+a_1, a+a_2, a+(b-a) = b, a+a_1+a_2, a+a_1+(b-a) = b+a_1, a+a_2+(b-a) = b+a_2, a+a_1+a_2+(b-a) = b+a_1+a_2\} = Q(a; a_1, a_2) \cup Q(b; a_1, a_2)$ implies that the three-cube $Q(a; a_1, a_2, b-a)$ is χ-monochromatic.

7.4 CHAPTER 5: THE HALES–JEWETT THEOREM

Exercise 5.1 The set of all roots in A_*^2 is $\{*a, *b, *c, *d, a*, b*, c*, d*, **\}$. It follows that all combinatorial lines A^2 are

$a\ a$	$a\ b$	$a\ c$	$a\ d$	$a\ a$	$b\ a$	$c\ a$	$d\ a$	$a\ a$
$b\ a$	$b\ b$	$b\ c$	$b\ d$	$a\ b$	$b\ b$	$c\ b$	$d\ b$	$b\ b$
$c\ a$	$c\ b$	$c\ c$	$c\ d$	$a\ c$	$b\ c$	$c\ c$	$d\ c$	$c\ c$
$d\ a$	$d\ b$	$d\ c$	$d\ d$	$a\ d$	$b\ d$	$c\ d$	$d\ d$	$d\ d$

□

Exercise 5.2 Observe that the line L_2 corresponds to $\{1\ 2\ 4, 2\ 2\ 3, 3\ 2\ 2, 4\ 2\ 1\}$. Since the first and third symbol in each word change to different symbols at different times, this set of words cannot be obtained from a root. It follows that L_2 does not correspond to a combinatorial line.

The remaining three Euclidean lines correspond to combinatorial lines:

Root	Combinatorial line	Euclidean line
$\tau = *32$	$L_\tau = \{1\ 3\ 2, 2\ 3\ 2, 3\ 3\ 2, 4\ 3\ 2\}$	L_1
$\sigma = ***$	$L_\sigma = \{1\ 1\ 1, 2\ 2\ 2, 3\ 3\ 3, 4\ 4\ 4\}$	L_3
$\theta = 24*$	$L_\theta = \{2\ 4\ 1, 2\ 4\ 2, 2\ 4\ 3, 2\ 4\ 4\}$	L_4

□

Exercise 5.3 For example, consider the following five roots:

$$\alpha = 1 * * * *, \beta = 1\ 5 * * * , \gamma = 1\ 5\ 5 * *, \ \delta = 1\ 5\ 5\ 5 *, \varepsilon = 1 * 5 * *.$$

The corresponding combinatorial lines are given by:

$L^{(\alpha)}$	$L^{(\beta)}$	$L^{(\gamma)}$	$L^{(\delta)}$	$L^{(\varepsilon)}$
1 1 1 1 1	1 5 1 1 1	1 5 5 1 1	1 5 5 5 1	1 1 5 1 1
1 2 2 2 2	1 5 2 2 2	1 5 5 2 2	1 5 5 5 2	1 2 5 2 2
1 3 3 3 3	1 5 3 3 3	1 5 5 3 3	1 5 5 5 3	1 3 5 3 3
1 4 4 4 4	1 5 4 4 4	1 5 5 4 4	1 5 5 5 4	1 4 5 4 4
1 5 5 5 5	1 5 5 5 5	1 5 5 5 5	1 5 5 5 5	1 5 5 5 5

Since $L_5^{(\alpha)} = L_5^{(\beta)} = L_5^{(\gamma)} = L_5^{(\delta)} = L_5^{(\varepsilon)} = 1\ 5\ 5\ 5\ 5$, the combinatorial lines $L^{(\alpha)}, L^{(\beta)}, L^{(\gamma)}, L^{(\delta)}$, and $L^{(\varepsilon)}$ are focused with the focus 1 5 5 5 5. □

Exercise 5.4 Here is a solution by Leo Moser from 1948.

Consider the "cube" of side k inside a cube of side $k + 2$. Clearly, every win will determine exactly one pair of surface elements, while each surface element determines exactly one win. Hence the number of wins will be half of the number of surface elements, which is the result stated [83]. □

Exercise 5.5 Observe that the number of combinatorial lines equals the number of roots in $(A \cup \{*\})^n$.

The number of all words of length n on the alphabet $A \cup \{*\}$ is $(m+1)^n$. Since the number of all words of length n on the alphabet A is m^n it follows that (# of combinatorial lines) = (# of roots in $(A \cup \{*\})^n$) = $(m+1)^n - m^n$. □

Exercise 5.6 There is no monochromatic combinatorial line in this red/black colouring of the cube A^2:

$*a$	$*b$	$*c$	$*d$	$a*$	$b*$	$c*$	$d*$	$**$
a a	$a\,b$	$a\,c$	$a\,d$	**a a**	$b\,a$	$c\,a$	d **a**	**a a**
$b\,a$	**b b**	$b\,c$	$b\,d$	$a\,b$	**b b**	$c\,b$	$d\,b$	**b b**
$c\,a$	$c\,b$	**c c**	$c\,d$	$a\,c$	$b\,c$	**c c**	$d\,c$	**c c**
d a	$d\,b$	$d\,c$	$d\,d$	$a\,d$	$b\,d$	**c d**	$d\,d$	$d\,d$

The existence of such a colouring does not contradict the Hales–Jewett theorem. It just shows that $HJ(2;4) > 2$. □

Exercise 5.7 There is no monochromatic combinatorial line in this colouring. Hence, $HJ(2;3) > 2$. *Note:* Observe that $\{(1,3,1),(2,2,1),(3,1,1)\}$ is a monochromatic set that contains three points that are collinear in $\mathbb{R}^3$. Does this set correspond to a combinatorial line in A^3?

Exercise 5.8 Let $r \geq 2$ and let $A = \{0,1\}$ be an alphabet. Let $2 \leq n < r$ and let $c : A^n \to [0, r-1]$ be an r-colouring defined in the following way: If $w = a_1 a_2 \cdots a_n$ then $c(w) = |\{i \in [1,n] : a_i = 0\}|$. Let $\tau \in A_*^n$ be any root. From $c(\tau_0) \neq c(\tau_1)$, it follows that the combinatorial line L_τ is not c-monochromatic. Hence, $HJ(r;2) \geq r$.

Let $n \geq r$. For $i \in [1, n+1]$, let the word $w_i = a_1^{(i)} a_2^{(i)} \dots a_n^{(i)}$ be such that $a_j^{(i)} = 0$, if $j < i$, and $a_j^{(i)} = 1$, if $j \geq i$. Let c be any r-colouring of the cube A^n. By the pigeonhole principle, there are $p, q \in [1, n+1]$, $p < q$, such that $c(w_p) = c(w_q)$. We consider the root $\tau = \underbrace{0\,0\dots 0}_{p-1} \underbrace{*\,*\dots *}_{q-p} \underbrace{1\,1\dots 1}_{n-q+1}$ and observe that the combinatorial line $L_\tau = \{\tau_0 = w_q, \tau_1 = w_p\}$ is c-monochromatic.

Note: Only in 2014, Hindman and Eric Tressler, an American mathematician and computer scientist, established that $HJ(2;3) = 4$ and obtained, what they called, "the first non-trivial Hales–Jewett number" [56]. □

Exercise 5.9 Let $N = HJ(2;\ell)$ and let c be a two-colouring of the set of natural numbers.

We define a two-colouring of the N-cube $[1,\ell]^N$ as follows: for $x_1 \; x_2 \; \cdots \; x_N \in [1,\ell]^N$, $c'(x_1 \; x_2 \; \cdots \; x_N) = c\left(2^{x_1} \cdot 2^{x_2} \cdot \ldots \cdot 2^{x_N}\right)$.

By the Hales–Jewett theorem, there is a c'-monochromatic line L_τ, determined by a root $\tau = a_1 \; a_2 \; \cdots \; a_N \in [1,\ell]^N_*$. Let $S = \{i \in [1,N] : a_i \in [1,\ell]\}$. Let $a = \sum_{i \in S} a_i$ and let $d = |[1,N]\backslash S|$.

Recall that $L_\tau = \{\tau_1, \tau_2, \ldots, \tau_\ell\} \subseteq [1,\ell]^N$, where, for $j \in [1,\ell]$, $\tau_j = a_1^{(j)} \; a_2^{(j)} \; \cdots \; a_N^{(j)}$, with $a_i^{(j)} = a_i$, if $i \in S$, and $a_i^{(j)} = j$, if $i \notin S$.

Note that, for any $j \in [1,\ell]$,

$$\sum_{i=1}^{N} a_i^{(j)} = \sum_{i \in S} a_i^{(j)} + \sum_{i \in [1,N]\backslash S} a_i^{(j)} = a + \sum_{i \in [1,N]\backslash S} j = a + jd.$$

On the other hand, $c'(\tau_1) = c'(\tau_2) = \cdots = c'(\tau_\ell)$, which together with $c'(\tau_j) = c\left(2^{\sum_{i=1}^{N} a_i^{(j)}}\right) = c\left(2^{a+jd}\right)$, for each $j \in [1,\ell]$, implies that the ℓ-term arithmetic progression $a_1 = a + d, a_2 = a + 2d, \ldots, a_\ell = a + \ell d$ has the required property. □

Exercise 5.10 Let c be an r-colouring of A and let $N = HJ(r;\ell)$. Let $x \in A$ be fixed.

We define an r-colouring of the N-cube $[1,\ell]^N$ as follows: for $n_1 \; n_2 \; \cdots \; n_N \in [1,\ell]^N$, $c'(n_1 \; n_2 \; \cdots \; n_N) = c\left(x^{n_1+n_2+\ldots+n_N}\right)$.

By the Hales–Jewett theorem, there is a c'-monochromatic line L_τ, determined by a root $\tau = a_1 \; a_2 \; \cdots \; a_N \in [1,\ell]^N_*$. Let $S = \{i \in [1,N] : a_i \in [1,\ell]\}$. Let $m = \sum_{i \in S} a_i$ and let $d = |[1,N]\backslash S|$.

Recall that $L_\tau = \{\tau_1, \tau_2, \ldots, \tau_\ell\} \subseteq [1,\ell]^N$, where, for $j \in [1,l]$, $\tau_j = a_1^{(j)} \; a_2^{(j)} \; \cdots \; a_N^{(j)}$, with $a_i^{(j)} = a_i$, if $i \in S$, and $a_i^{(j)} = j$, if $i \notin S$.

Note that for any $j \in [1,\ell]$,

$$\sum_{i=1}^{N} a_i^{(j)} = \sum_{i \in S} a_i^{(j)} + \sum_{i \in [1,N]\backslash S} a_i^{(j)} = a + \sum_{i \in [1,N]\backslash S} j = a + jd.$$

On the other hand $c'(\tau_1) = c'(\tau_2) = \cdots = c'(\tau_\ell)$ which implies that, for each $j \in [1,\ell]$, $c'(\tau_j) = c\left(x^{\sum_{i=1}^{N} \tau_j(i)}\right) = c\left(x^{m+jd}\right)$.

Observe that, since binary operation $\bullet$ is associative, it follows, for each $j \in [1,\ell]$,

$$
\begin{aligned}
x^{m+jd} &= \underbrace{x \bullet x \bullet \ldots \bullet x}_{m+jd} = \underbrace{(x \bullet \ldots \bullet x)}_{m+d} \bullet \underbrace{(x \bullet \ldots \bullet x)}_{(j-1)d} \\
&= \underbrace{(x \bullet \ldots \bullet x)}_{m+d} \bullet \underbrace{((\underbrace{x \bullet \ldots \bullet x}_{d}) \bullet \ldots \bullet (\underbrace{x \bullet \ldots \bullet x}_{d}))}_{j-1} = x^{m+d} \cdot \left(x^d\right)^{j-1}.
\end{aligned}
$$

Hence, for $a = x^{m+d}$ and $b = x^d$, the set $\{a, a \bullet b, a \bullet b^2, \ldots, a \bullet b^{\ell-1}\}$ is c-monochromatic.

Note: Observe that, in the case of the semigroup $(\mathbb{N}, +)$, Gallai's theorem for semigroups becomes van der Waerden's theorem. Also, observe that the set of all non-negative integer powers of 2, with the usual multiplication, is a semigroup. □

Exercise 5.11 Let $r \in \mathbb{N}$ and let $N = HJ(r; m)$. Let c be an r-colouring of the set V.

We define c', an r-colouring of the N-cube A^N, by $c'(x_1 \ x_2 \ \cdots \ x_N) = c(x_1 + x_2 + \ldots + x_N)$, $x_1 \ x_2 \ \cdots \ x_N \in A^N$.

By the Hales–Jewett theorem, there is a c'-monochromatic line L_τ, determined by a root $\tau = v_1 \ v_2 \ \cdots \ v_N \in A_*^N$. Let $S = \{i \in [1, N] : v_i \in A\}$. Let $u = \sum_{i \in S} v_i \in V$ and let $\lambda = |[1, N] \backslash S|$.

Recall that $L_\tau = \{\tau_{a_1}, \tau_{a_2}, \ldots, \tau_{a_m}\} \subseteq A^N$, where, for $j \in [1, m]$, $\tau_{a_j} = v_1^{(j)} \ v_2^{(j)} \ \cdots \ v_N^{(j)}$, with $v_i^{(j)} = v_i$, if $i \in S$, and $v_i^{(j)} = a_j$, if $i \notin S$.

Note that for any $j \in [1, m]$,

$$\sum_{i=1}^N v_i^{(j)} = \sum_{i \in S} v_i + \sum_{i \in [1, N] \backslash S} a_j = u + \lambda \cdot a_j.$$

On the other hand, $c'(\tau_{a_1}) = c'(\tau_{a_2}) = \cdots = c'(\tau_{a_m})$, which implies that

$$c'(\tau_{a_j}) = c\left(\sum_{i=1}^N \tau_{a_j}(i)\right) = c\left(u + \lambda \cdot a_j\right), \text{ for each } j \in [1, m].$$

Therefore the set $u + \lambda A = \{u + \lambda \cdot a_1, u + \lambda \cdot a_2, \ldots, u + \lambda \cdot a_m\}$ is c-monochromatic. □

7.5 CHAPTER 6: HAPPY END PROBLEM

Exercise 6.1 We consider three cases:

Case 1: Suppose that the five vertices form a convex pentagon. Then any four of them form a convex quadrilateral. There are $\binom{5}{4} = 5$ convex quadrilaterals in this case:

Case 2: Suppose that four points form a convex quadrilateral (say A_1, A_2, A_3, and A_4 in clockwise order) which contains the fifth point A_5 in its interior. Let S be the intersection of the diagonals $\overline{A_1 A_3}$ and $\overline{A_2 A_4}$. Then the point A_5 lies in one of the four triangles into which diagonals dissect the quadrilateral, say $\triangle S A_1 A_2$. Observe that the quadrilaterals $A_1 A_5 A_3 A_4$ and $A_2 A_3 A_4 A_5$

are convex but $A_1A_2A_3A_5$ and $A_1A_2A_4A_5$ are not. There are three convex quadrilaterals in this case.

Case 3: Suppose that three points (say A_1, A_2, A_3) form a triangle with A_4 and A_5 in its interior. The line A_4A_5 intersects two of the three sides, say $\overline{A_1A_2}$ and $\overline{A_1A_3}$. Then $A_2, A_3A_4A_5$ is the only convex quadrilateral. □

Exercise 6.2

Consider the $(4,4,2)$ configuration on the figure to the right. As we already know, the eight points in the $(4,4)$ part of the given configuration do not contain a convex pentagon.

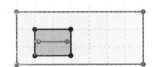

Suppose that there is a convex hexagon in this configuration. Then the hexagon has to contain both green points inside of the blue quadrilateral. (If the convex hexagon contains only one green point then the remaining five points belong to the set of red and blue points, which contradicts the fact that those points do not contain a convex pentagon.)

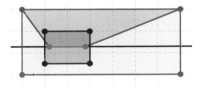

Since the line determined by the two green points divides the plane in the way that each of the two half-planes contains only two red and two blue points, the convex hexagon must have two red, two blue, and two green vertices. But this hexagon is not convex because the blue points are inside of the quadrilateral determined by the red and green points. □

Exercise 6.3 By Esther Klein's result we know that any five points in the plane in general position contain a convex quadrilateral.

If the given five points are vertices of a convex pentagon then any four points form an empty convex quadrilateral.

Say that the given five points S do not determine a convex pentagon. Then the convex hull of the set S has three or four points. In both cases there is an empty convex quadrilateral.

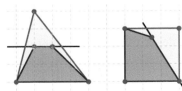

Let S be a set of n points in the plane in general position, with $n \geq 6$.

Let $L = \{l_i : i \in [1, n(n-1)/2]\}$ be the set of all lines in the plane determined by pairs of points from the set S. Let p be a line that is not parallel to any line from the set L and such that all points that belong to the set S lie on the same side of the line p. Observe that this implies that none of the line segments with the end points from S intersects the line p.

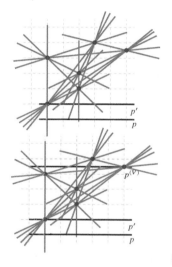

Move the line p towards the set S. Let p' be the line that contains a point from S, that is parallel to the line p, and is such that there is no element from S between p and p'.

Continue moving the line p till it reaches five points from S. Note that the line $p^{(V)}$ divides the set S so that one point from S is on the $p^{(V)}$, only four points from S are below $p^{(V)}$, and all other points from S are above $p^{(V)}$.

What should we do next? ☐

Exercise 6.4 By Ramsey's theorem there is a set T, $T \subset S$, and $|T| = n$, such that for any $i, j, k \in T$, $i < j < k$, the set $\{i, j, k\}$ is always of the same colour, say blue. But this implies that any four points from T form a convex quadrilateral. To see this take $x, y, z, w \in T$ and suppose that x, y, z, and w determine a concave quadrilateral. Suppose that the point w is inside of the triangle determined by x, y, and z. Also, suppose that $x < y < z$. Let the triple $\{x, y, z\}$ be coloured blue.

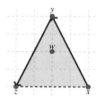

Each of four possible cases leads to a contradiction: a three-element subset of T that is coloured red:

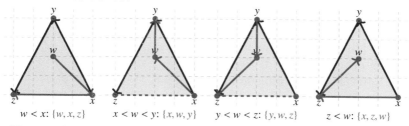

$w < x$: $\{w, x, z\}$ $x < w < y$: $\{x, w, y\}$ $y < w < z$: $\{y, w, z\}$ $z < w$: $\{x, z, w\}$

Observe that Tarsi's proof establishes that $ES(n) \leq R(3; n, n)$.   □

Exercise 6.5

Consider the configuration S of six points in the plane in general position shown in the figure to the right. Note that there is *no* a four-cup or a four-cap in this configuration.

Hence $f(4, 4) \geq 7$, which implies that $f(4, 4) = 7$.

Let A be a configuration of six points with no four-cup or four-cap. Let B be a four-cup.

We place configurations A and B in the way that the following conditions are satisfied: (a) Every point in B has greater first coordinate than any point in A; and (b) The slope of any line connecting a point in A with a point in B is greater than the slope of any line connecting two points in the same configuration. Let $S = A \cup B$.

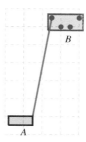

We observe that any cup in S that contains both points from A and B can contain only *one* point from B. Therefore the configuration S contains at most a four-cup.

Clearly, the configuration S does not contain a four-cap.

Hence, the configuration S contains neither a five-cup nor a four-cap. It follows that $f(5, 4) \geq |S| + 1 = |A| + |B| + 1 = 6 + 4 + 1 = 11$, which implies $f(5, 4) = f(4, 5) = 11$.

Finally we consider two configurations, X and Y, that satisfy the following conditions:
(a) $|X| = |Y| = 10$; (b) The configuration X does not contain a four-cup or a five-cap; (c) The configuration Y does not contain a five-cup or a four-cap; (d) Every point in Y has greater first coordinate than any point in X; and (e) The slope of any line connecting a point in X with a point in Y is greater than the slope of any line connecting two points in the same configuration. Let $Z = X \cup Y$.

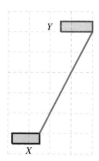

Observe that: (a) Any cup in Z that contains points from both X and Y can contain only *one* point from Y. Therefore the configuration Z can contain at most a four-cup; and (b) Any cap in Z that contains points from both X and Y can contain only *one* point from X. The configuration Z can contain at most a four-cap.

Hence, the configuration S contains neither a five-cup nor a five-cap. It follows that $f(5,5) \geq |Z| + 1 = |X| + |Y| + 1 = 10 + 10 + 1 = 21$, and $f(5,5) = 21$. □

Exercise 6.6 Let $N = \binom{k+l-4}{k+2}$ and let $R_{k,l} = \{(x_i, y_i) : i \in [1, N]\}$ be a set of points in general position with no k-cups or l-caps and such that $i < j$ implies that $x_i < x_j$.

Let $\eta \in \left(0, \varepsilon \cdot \min\left\{ \left| \frac{x_j - x_i}{y_j - y_i} \right| : i, j \in [1, N], y_i \neq y_j \right\}\right)$. Let $T_{k,l} = \{(x_i, \eta y_i) :$ $i \in [1, N]\}$. From, for any $i, j \in [1, N]$, $y_i \neq y_j$, $\left| \frac{\eta y_j - \eta y_i}{x_j - x_i} \right| < \varepsilon \cdot \left| \frac{x_j - x_i}{y_j - y_i} \right| \cdot \left| \frac{y_j - y_i}{x_j - x_i} \right| = \varepsilon$, it follows that each line that contains two points of $T_{k,l}$ is with the slope whose absolute value is less than ε.

Observe that, since $\eta > 0$, the configuration of cups and caps in $T_{k,l}$ is the same as the configuration of cups and caps in $R_{k,l}$. Hence, $T_{k,l}$ is a set of N points in general position with no k–cups or l–caps.

Let $d = \max\left\{ \sqrt{(x_i - x_j)^2 + \eta^2(y_i - y_j)^2} : i, j \in [1, N] \right\}$, i.e. let d be the diameter of the set $T_{k,l}$.

Let $s = \frac{r}{d}$ and let $F_{k,l} = \{(sx_i, s\eta y_i) : i \in [1, N]\}$. Then, for $i \neq j$, $\left| \frac{s\eta y_j - s\eta y_i}{sx_j - sx_i} \right| = \eta \cdot \left| \frac{y_j - y_i}{x_j - x_i} \right| < \varepsilon$. Also, $(sx_i - sx_j)^2 + (s\eta y_i - s\eta y_j)^2 \leq s^2 d^2 = r^2$.

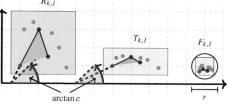

Hence, the set $F_{k,l}$ is with the required properties. □

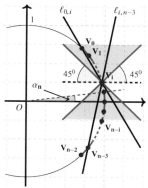

Exercise 6.7 Let $\alpha_n = \frac{\pi}{2(n-2)}$ and let, for $i \in [0, n-2]$, $V_i = \left(\cos\left(\frac{\pi}{4} - i\alpha_n\right), \sin\left(\frac{\pi}{4} - i\alpha_n\right)\right)$. Note that $\angle V_{i+1}OV_i = \alpha_n$, for each $i \in [0, n-3]$. Observe that, for any $i, j \in [0, n-2]$, $i < j$, the absolute value of the slope (if it exists) of the line $\ell_{i,j}$ through the points V_i and V_j is greater than 1.

For each $i \in [0, n-2]$, we construct a circle C_i with centre at V_i and diameter r. We choose $r > 0$ so that all circles are mutually disjoint. In addition, the radius r is such that the absolute value of the slope of any non-vertical line that intersects both C_i and C_j is greater than 1 and that no line intersects three different circles.[1]

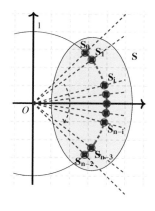

Inside each circle C_i we place a set S_i, where $S_0 = \{V_0\}$, $S_{n-2} = \{V_{n-2}\}$, and, for $i \in [1, n-3]$, S_i is a set of $\binom{n-2}{i}$ points in general position, with no $(n-i)$-cups or $(i+2)$-caps, and such that each line that contains two points of S_i is with the slope whose absolute value is less than 1. Let $S = \bigcup_{i=0}^{n-2} S_i$. Since, for $i \neq j$, $S_i \cap S_j = \emptyset$, it follows that $|S| = \sum_{i=0}^{n-2} |S_i|$. Recall that $|S_0| = 1 = \binom{n-2}{0}$, $|S_{n-2}| = 1 = \binom{n-2}{n-2}$, and, for $i \in [1, n-3]$, $|S_i| = |F_{i+2,n-i}| = \binom{n-2}{i}$. Therefore, $|S| = \sum_{i=0}^{n-2} \binom{n-2}{i} = 2^{n-2}$.

Observe that our choice of r guarantees that there are no three collinear points in S.

Let m be a positive integer and let $\Pi = P_1 P_2 \cdots P_m$ be a convex m–gon, oriented clockwise, with $P_i \in S$, for all $i \in [1, m]$. Let k be the least $i \in [0, n-3]$

---

[1] We can use calculus to establish that such $r > 0$ exists, i.e. that the value of the slope of any non-vertical line that intersects both C_i and C_j is approximately equal to the slope of $\ell_{i,j}$. Fore example, let $A = (a,b)$, $B = (c,d)$, $A' = (a + \eta_1, b + \eta_2)$, and $B' = (c + \sigma_1, d + \sigma_2)$ be points such that $a \neq c$ and $\eta_1, \eta_2, \sigma_1, \sigma_2 \in (-r, r)$. Then, for $a - c = m$, $b - d = n$, $\eta_1 - \sigma_1 = \delta_1$, and $\eta_2 - \sigma_2 = \delta_2$, the slope of the line through A and B is $\frac{n}{m}$ and the slope of the line through A' and B' is $\frac{n+\delta_2}{m+\delta_1}$. It follows that $\left|\frac{n+\delta_2}{m+\delta_1} - \frac{n}{m}\right| = \left|\frac{n\delta_1 - m\delta_2}{m^2 + m\delta_1}\right| \to 0$, as $r \to 0$.

among all those S_i's that contain a vertex of the polygon Π. Let P_1 be the vertex with the smallest x-coordinate in S_k.

Suppose that all vertices of the polygon Π belong to S_k. Let $s \in [2, m]$ be such that P_s has the largest x-coordinate among all vertices of Π. But then P_1 and P_s are the end points of a cup or a cup or a cup *and* a cap contained in S_k.

Recall that in S_k all cups are of the size at most $k - 1$ and all caps are of the size at most $n - k - 1$. By the inclusion-exclusion principle and because P_1 and P_s are possibly double counted, it follows that $m \le (k + 1) + (n - k - 1) - 2 = n - 2$. Therefore no S_i, $i \in [1, n - 2]$, contains a convex n-gon.

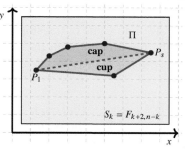

Suppose that there is $i \in [2, m]$ such that $P_i \notin S_k$. Let l be the largest among all $i \in [k + 1, n - 2]$ for which S_i contains a vertex of Π.

Under this assumption, the vertices of Π contained in S_k do not form a cup in S_k. Otherwise, by the construction of the set S, the line segment $\overline{P_1 P_s}$ would be outside of the polygon Π. It follows that S_k contains at most $k + 1$ (which is the size of the largest possible cap in S_k) vertices of the polygon Π.

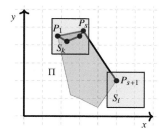

Similarly, the vertices of Π contained in S_l do not form a cap in S_l. Hence, S_l contains at most $n - l - 1$ (which is the size of the largest possible cup in S_l) vertices of the polygon Π.

Finally, suppose that $i \in [k + 1, l - 1]$ is such that S_i contains a vertex of Π.

Suppose that S_i contains more than one vertex of Π. Let $P_t \in S_i$ be such that $P_{t-1} \notin S_i$. Let p_t be the line through P_{t-1} and P_t. Recall that the absolute value of the slope of the line p_t is greater than 1. Also, recall that the absolute value of the slope of any line that contains two points from S_i is less than 1. We observe that, since Π is convex, all of its vertices, except P_{t-1} and P_t are on the same side of the line p_t.

Next, we draw lines, q and r, with slopes ± 1 through the point P_t. In addition, we observe that any vertex of Π, other than P_t that belongs to S_i must belong to one of the two sectors determined by the lines q and r that does not contain the line p_t.

Let $s > t + 1$ be such that the vertex P_s does not belong to S_i but that the vertex $P_{s-1} \in S_i$. Since the slope of the line through P_t and P_s is greater than 1, the point P_s belongs to the region that is bounded by rays that belong to the lines r (or q) and p_t. Since none of the vertices of Π between P_t and P_s belongs to this region, it follows that the line segment $\overline{P_t P_s}$ is not contained in Π. This contradicts our assumption that Π is convex.

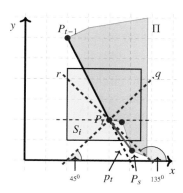

Therefore, for each $i \in [k+1, l-1]$, S_i contains at most one vertex of Π.

We have established that each vertex of Π is one of the most $k+1$ vertices of Π that belong to S_k or it is one of the most $n-l-1$ vertices of Π that belong to S_l or it is the unique vertex that belongs to S_i for some $i \in [k+1, l-1]$. This implies that $m \leq (k+1) + (n-l-1) + ((l-1) - (k+1) + 1) = n - 1$.

It follows that the set S does not contain vertices of a convex n-gon. □

Exercise 6.8

1. Observe that $\triangle ABG$ and $\triangle GBC$ are two equilateral triangles that share the side $\overline{BG}$. This implies that the points A and C are symmetric with the respect to the line BG. Hence, $|\overline{AC}| = \sqrt{3}$. Similarly, $|\overline{AD}| = \sqrt{3}$.

2. By the Law of Cosines, from $\triangle ACD$, it follows that $\cos(\angle CAD) = \frac{|\overline{AC}|^2 + |\overline{AD}|^2 - |\overline{DC}|^2}{2|\overline{AC}| \cdot |\overline{AD}|} = \frac{5}{6}$ and $\alpha = m(\angle CAD) = \arccos\left(\frac{5}{6}\right) \approx 33.56^0$.

   Let ρ be the clockwise rotation with the centre of rotation at the point A and the angle of rotation equal to α. Observe that $\rho(A) = A$ and $\rho(D) = C$. Since ρ is an isometry it follows that $|\overline{AF}| = |\overline{FD}| = 1 \Rightarrow |\rho(A)\rho(F)| = |\rho(F)\rho(D)| = |A\rho(F)| = |\rho(F)C| = 1$.

   Since B and G are the only two points in the plane at a unit distance from both A and C and because of the direction of the rotation, it follows that $\rho(F) = B$. By using the fact that ρ is an isometry again, we conclude that $\rho(E) = G$.

   Therefore, $m(\angle EAG) = m(\angle FAB) = \alpha$. Let $x = m(\angle FAG)$. Then $m(\angle EAB) = m(\angle EAG) + m(\angle GAF) + m(\angle FAB) = 2\alpha + x$.

   On the other hand, $m(\angle EAB) = m(\angle EAF) + m(\angle GAB) - m(\angle GAF) = 2 \cdot 60^0 - x = 120^0 - x$.

   It follows that $2\alpha + x = 120^0 - x$ which implies $x = 60^0 - \alpha \approx 26.44^0$.

3. Start by choosing a point A in the plane and then draw a circle with the centre at A and radius 1.

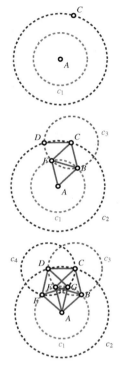

Denote this circle by c_1. Next, draw a circle with the centre at A and radius $\sqrt{3}$. Denote this circle by c_2. Choose a point C on the circle c_2.

Draw a circle with the centre at C and radius 1. Denote this circle by c_3. Let D be a point of intersection of c_2 and c_3. Let B and F be the points of intersection of c_1 and c_3. Draw the line segments $\overline{AB}$, $\overline{AF}$, $\overline{BC}$, $\overline{BF}$, $\overline{CD}$, and $\overline{CF}$. Observe that those line segments are of length 1. How would you argue that $|\overline{BF}| = 1$?

Draw a circle with the centre at D and radius 1. Denote this circle by c_4. Let E and G be the points of intersection of c_1 and c_4. Draw the line segments $\overline{AE}$, $\overline{AG}$, $\overline{DE}$, $\overline{DG}$, and $\overline{EG}$. Observe that those line segments are of length 1. The Moser spindle appears!

□

Exercise 6.9

1. Consider an equilateral triangle and use the pigeonhole principle.

2. Observe that the existence of four points that are friends with each other would imply the existence of a triangle inscribed in the unit circle with all its sides equal to 1, which is impossible. A mutual friend of two friends, say A and B, must belong to the intersection of circles with their centres at A and B and radii equal to 1. Hence, two points can have at most two common friends.

3. Say that *A* and *B* are friends. Use two colours to colour those two points. Suppose that they have two mutual friends, *C* ad *D*. Use the third colour to colour *C*. How should we colour the point *D*?

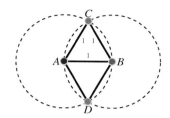

4. We consider a set $S = \{A, B, C, D, E\}$ of five points in the plane. Observe that not each of the points *A*, *B*, *C*, *D*, and *E* can have exactly three friends from *S*. Otherwise, if, say, *A* and *E* are not friends, then *B*, *C*, and *D* would be their common friends, which is impossible by Part 2.

Suppose that the point *E* does not have three friends. Say that *E* has two friends, *B* and *C*. Use Part 3 to properly three-colour points *A*, *B*, *C*, and *D*. How should we colour the point *E*?

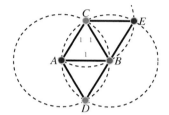

If *A* has four friends then those four points belong to a circle with the centre at *A* and we can colour them properly with three colours.

5. Let *A*, *B*, *C*, *D*, *E*, and *F* be six points in the plane.

If there is a point with two or four or five friends then we can use a similar argument as in Part 4.

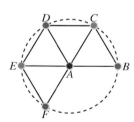

Suppose that each point has exactly three friends. Say that *A* is a friend with *B*, *C*, and *D*. Since *A* and *E* are not friends, three friends of *E* are among *B*, *C*, *D*, and *F*. Since *A* and *E* can have at most two mutual friends, say *C* and *D*, we conclude that *E* and *F* are friends.

The remaining two friends of F are among B, C, and D. By the pigeonhole principle, at least one of C or D must be a friend of F. Say, D. Observe that F cannot be a friend with both C and D, because that would imply that C and D have three mutual friends, A, E, and F, what is impossible.

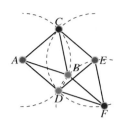

6. Observe that a proper colouring of the Moser spindle requires four colours.   □

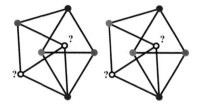

Exercise 6.10

1. If both (A, C) and (A, D) are monochromatic then the pair (C, D) is monochromatic as well. But, this has to be avoided.

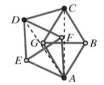

2. Observe that the two spindles share four vertices and five edges.

This is what de Grey meant when saying "interlocking spindles." Also note that the outer polygon is a hexagon of side-length 1.

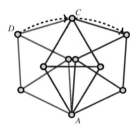

3. Say that A is the vertex of degree 6. Colour A blue and then colour the vertices not adjacent to A by yellow and green alternatively. Use blue and red to colour the remaining vertices. To write a program you may use the following strategy: Denote the four colours by the numbers 1, 2, 3, and 4. Use 0 to denote the vertex that is not coloured. Then go through each vertex that was not coloured, get all the numbers of the vertices around it, then colour the vertex by the lowest positive number that is not in that set. For example, if a vertex had vertices around it with the numbers 0, 1, 2, and 2, then this vertex would have to be given the number 3 (which means colour number three).

4. Use the same idea as above to find a proper four-colouring by hand.

Alternatively, you may write a program that generates proper four-colourings. Here is a four-colouring generated by a Python program written by Ewan Brinkman, a first-year student at Simon Fraser University.

Notice, what de Grey calls, "monochromatic $\sqrt{3}$-apart vertex pairs (...) in any four-colouring." □

Exercise 6.11

1. From $\triangle XDC$ it follows that $|\overline{XD}| = |\overline{XC}| \cdot \cos\theta = \frac{\cos\theta}{2}$. Hence, $|\overline{XY}| = |\overline{XD}| + |\overline{DE}| + |\overline{EY}| = \frac{\cos\theta}{2} + 1 + \frac{\cos\theta}{2} = 1 + \cos\theta$.

2. We distinguish two cases:

Case 1: There is a vertex X in the lattice such that M and N belong to the same lump or that they belong to two different lumps that are associated to X (i.e M and N are in two different triangles in the lattice that share the vertex X).
In this case M and N belong to the interior of a circle with radius equal to $\frac{1}{2}$ which implies that $|\overline{MN}| < 1$.

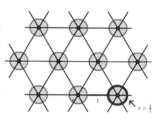

Case 2: There are two vertices in the lattice, X and X', such that M belongs to a lump centred at X and N belongs to a lump centred at X'.

By construction, the distance between a point that belongs to the boundary of the lump centred at X and a point that belongs to the boundary of the lump centred at X' is greater than or equal to 1. Hence, in this case, $|\overline{MN}| > 1$.

3. Consider the lump centred at the vertex X and observe that $\mathcal{A} = $ (Area of $\triangle XDC$) + (Area of circular sector XBC) + (Area of $\triangle XAB$).

Next, from (Area of $\triangle XDC$) = (Area of $\triangle XAB$) = $\frac{1}{2} \cdot \left(\frac{1}{2} \cdot \sin\theta\right) \cdot \left(\frac{1}{2} \cdot \cos\theta\right) = \frac{\sin(2\theta)}{16}$ and (Area of circular sector XBC) = $\frac{1}{2} \cdot \left(\frac{1}{2}\right)^2 \cdot \left(\frac{\pi}{3} - 2\theta\right) = \frac{1}{8} \cdot \left(\frac{\pi}{3} - 2\theta\right)$, it follows that $\mathcal{A} = \mathcal{A}(\theta) = \frac{1}{8} \cdot \left(\sin(2\theta) + \frac{\pi}{3} - 2\theta\right)$.

4. Since the side of the equilateral triangle is $1 + \cos\theta$, it follows that
$A_\triangle = \frac{\sqrt{3}\cdot(1+\cos\theta)^2}{4}$.

5. Recall that the question is to determine the portion of the plane that the set $S = S(\theta)$ occupies. Because all triangles in the lattice are congruent, the density of the set S in the plane will be equal to the portion of a single triangle in the lattice that the three lumps occupy. Hence,
density of $S(\theta) = \delta(S(\theta)) = \frac{3\cdot\mathcal{A}}{A_\triangle} = \frac{\sqrt{3}}{2}\cdot\frac{\sin(2\theta)+\frac{\pi}{3}-2\theta}{(1+\cos\theta)^2}$.

6. From $\frac{d\delta}{d\theta} = 2\sqrt{3}\cdot\frac{\sin\theta\cdot(\frac{\pi}{6}-\sin\theta-\theta)}{(1+\cos\theta)^3}$, it follows that $\frac{d\delta}{d\theta} = 0$ if and only if $\sin\theta + \theta = \frac{\pi}{6}$. Observe that, by the Intermediate Value Theorem, there is $\theta_0 \in \left(0, \frac{\pi}{6}\right)$ such that $\sin\theta_0 + \theta_0 = \frac{\pi}{6}$. Technology gives, $\theta_0 \approx 0.263316$ radians.

To decide if the critical number $\theta = \theta_0$ yields the absolute maximum value of $\delta = \delta(S(\theta))$, we use the First Derivative Test.

Observe that the inequality $\frac{d\delta}{d\theta} < 0$ is equivalent to the inequality $\eta(\theta) = \frac{\pi}{6} - \sin\theta - \theta > 0$.

From $\frac{d\eta}{d\theta} = -\cos\theta - 1 < 0$, $\theta \in \left(0, \frac{\pi}{6}\right)$, it follows that the function $\eta(\theta)$ is decreasing from $\eta(0) = \frac{\pi}{6}$ to $\eta(\frac{\pi}{6}) = -\frac{1}{2}$. Also, this implies that θ_0 is the only critical number of the function δ.

Hence, when passing through $\theta = \theta_0$, $\frac{d\delta}{d\theta}$ changes its sign from positive to negative, and, by the First Derivative Test, the number $\delta = \delta(S(\theta_0))) \approx \delta(S(0.263316)) \approx 0.229365$ is the local and absolute maximum value of the function $\delta = \delta(S(\theta))$.

7. Let $\mathcal{T}$ be the family of all sets in the plane that avoid a unit distance. By definition, $m_1(\mathbb{R}^2) = \sup\{\delta(T) : T \in \mathcal{T}\}$. Since $S(\theta_0) \in \mathcal{T}$, it follows that $m_1(\mathbb{R}^2) \geq \delta(S(\theta_0)) > 0.2293$.

Exercise 6.12

1. Since $\mathcal{P}$ is a set of non-collinear points, it follows that $|\mathcal{L}| > 1$. Any line in $\mathcal{L}$ contains at least two different points from the set $\mathcal{P}$. This implies that $|\mathcal{L}| \leq \binom{n}{2} = \frac{n(n+1)}{2}$. It follows that $\mathcal{L} = \{\ell_1, \ell_1, \ldots, \ell_m\}$, where $2 \leq m \leq \frac{n(n+1)}{2}$.

2. Recall that $\mathcal{P}$ is a set of non-collinear points.

3. Observe that, for $i \in [1, m]$, since the set $\mathcal{P}_i$ is finite, the set $\{d_{i,A} : A \in \mathcal{P}_i\}$ is also finite. Hence, the set $D = \cup_{i=1}^{m}\{d_{i,A} : A \in \mathcal{P}_i\}$ is a finite set of

positive numbers and, as such, it has a minimum value $d > 0$. Since $d \in D$, there is at least one line, call it ℓ_{i_0}, and at least one point $A_{i_0} \in \mathcal{P}_{i_0}$ such that $d_{i_0, A_{i_0}} = d$.

4. Observe that, by definition, $|\overline{AX}| = d_{i,A}$.

Case 1: Suppose that $X \in \{K, M, N\}$, say, $X = K$. Let $\ell_j \in \mathcal{L}$ be the line passing through the points A and M. Suppose that $Y \in \ell_j$ is such that $\overline{XY} \perp \ell_j$. By definition $|\overline{XY}| = d_{j,X}$.

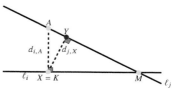

Since $\overline{AX}$ is the hypothenuse in the right triangle $\triangle AYX$, it follows that $d_{i,A} = |\overline{AX}| > |\overline{XY}| = d_{j,X} \geq d$.

Case 2: Suppose that $X \notin \{K, M, N\}$. By the pigeonhole principle, two points, say, N and M, are on the same side of X. Suppose that the point N is between X and M. Let $\ell_j \in \mathcal{L}$ be the line passing through the points A and M. Let $Y, Z \in \ell_j$ be such that $\overline{XY} \perp \ell_j$ and $\overline{NZ} \perp \ell_j$. By definition $|\overline{NZ}| = d_{j,N}$.

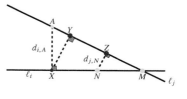

Since $\overline{AX}$ is the hypothenuse in the right triangle $\triangle AYX$, and since $\overline{NZ}$ is a line segment parallel to $\overline{XY}$ and inside $\triangle MYX$, it follows that $d_{i,A} = |\overline{AX}| > |\overline{XY}| > |\overline{NZ}| = d_{j,N} \geq d$.

5. The line ℓ_{i_0}, established in Part 3, is passing through exactly two points that belong to $\mathcal{P}$. Otherwise, by Part 4, $d > d$, which is impossible. □

Bibliography

[1] M. Aigner and G.M. Ziegler. *Proofs from THE BOOK*. Springer, 6th edition, 2018. Berlin, Germany.

[2] G.L. Alexanderson and P. Erdős. An interview with Paul Erdős. *The Two–Year College Mathematics Journal*, 12(4):249–259, 1981.

[3] P. G. Anderson. A generalization of Baudet's conjecture (van der Waerden's theorem) *American Mathematical Monthly* 83:359–361, 1976.

[4] J. Andrews, A. Blondal, T. Trinh, and H. Zhang. Quad-Tac-Toe game. In V. Jungić, editor. *Introduction to Ramsey Theory: Students' Projects*. Simon Fraser University, Burnaby, BC, 2022.

[5] L. Babai. Paul Erdős just left town. *Notices of The AMS*, 45(1):66–73, January 1998.

[6] T. Bang. On the sequence $[n\alpha]$, $n = 1, 2, \ldots$ – Supplementary note to the preceding paper by Th. Skolem. *Mathematical Scandinavian*, 5:69–76, 1957.

[7] J. Beck. Combinatorial Games: Tic-Tac-Toe Theory. Cambridge University Press, Cambridge, England, 2008.

[8] V. Bergelson and A. Leibman. Polynomial extensions of van der Waerden's and Szemerédi's theorems. *Journal of American Mathematical Society*, 9(3):725–753, 1996.

[9] T.F. Bloom and O. Sisask. Breaking the logarithmic barrier in Roth's theorem on arithmetic progressions. *arXiv*, July 2020.

[10] W.E. Bonnice. On convex polygons determined by a finite planar set. *American Mathematical Monthly*, 81:749–752, 1974.

[11] J.M. Borwein. The life of modern homo habilis mathematicus: Experimental computation and visual theorems. In J. Monaghan, L. Trouche,

and J.M. Borwein, editors, *Tools and Mathematics: Instruments for Learning*, Mathematics Education Library, pages 23–90. Springer, Cham, Switzerland, 2016.

[12] P. Borwein and W. Moser. A survey of Sylvester's problem and its generalizations. *Aequationes Mathematicae*, 40(1):111–135, 1990.

[13] T. C. Brown. Abstract 74T-A113. *Notices of the American Mathematical Society*. 21:A-432, 1974.

[14] T.C. Brown, R.L. Graham, and B. Landman. On the set of common differences in van der Waerden's theorem on arithmetic progressions. *Canadian Mathematical Bulletin*, 42:25–36, 1999.

[15] T.C. Brown and P.J.-S. Shiue. On the history of van der Waerden's theorem on arithmetic progressions. *Tamkang Journal of Mathematics*, 4:335–341, 2001.

[16] V. Chvátal. Some unknown van der Waerden numbers. In R. Guy, editor, *Combinatorial Structures and their Applications. Proceedings of the Calgary International Conference on Combinatorial Structures and Their Applications*, pages 31–33. Gordon and Breach, New York, 1970.

[17] D. Conlon. A new upper bound for diagonal Ramsey numbers. *Annals of Mathematics*, 2(170):941–960, 2009.

[18] M. Cowling. A world of teaching and numbers – times two. *The Sydney Morning Herald*, November 2005.

[19] H.T. Croft. Incidence incidents. *Eureka (Cambridge)*, 30:22–26, 1967.

[20] N.G. de Bruijn. Commentary. In E. M. J. Bertin, H. J. M. Bos, and A. W. Grootendorst, editors, *Two Decades of Mathematics in the Netherlands 1920–1940*, pages 16–124. Mathematical Centre, Amsterdam, 1976.

[21] A.D.N.J. de Grey. The chromatic number of the plane is at least 5. *Geombinatorics*, 28:5–18, 2018.

[22] W. Deuber. On van der Waerden's theorem on arithmetic progressions. *Journal of Combinatorial Theory, Series* 32:115–118, 1982.

[23] Y. Dold-Samplonius. Interview with Bartel Leendert van der Waerden. *Notices of The AMS*, 44(3):313–320, 1993.

[24] S. Eliahou and M.P. Revauelta. The Schur degree of additive sets. *Discrete Mathematics*, 344:112332, 2021.

[25] P. Erdős. Beweis eines satzes von tschebyschef. *Acta Litterarum ac Scientiarum Szeged*, 5:194–198, 1932.

[26] P. Erdős. Some remarks on the theory of graphs. *Bulletin of the American Mathematical Society*, 53(4):292–294, 1947.

[27] P. Erdős and R.L. Graham. *Old and New Problems and Results in Combinatorial Number Theory*. L'Enseignement Mathématique, Université de Genéve, Genéve, Switzerland, 1980.

[28] P. Erdős and G. Szekeres. A combinatorial problem in geometry. *Compositio Mathematica*, 2:463–470, 1935.

[29] P. Erdős and G. Szekeres. On some extremum problems in elementary geometry. *Annales Universitatis Scientiarum Budapestinensis de Rolando Eotvos Nominatae. Sectio Mathematica*, 3–4:53–62, 1960.

[30] P. Erdős. On some problems of elementary and combinatorial geometry. *Annali di Matematica pura ed applicata*, 103:99–108, 1975.

[31] P. Erdős. Combinatorial problems in geometry. *Mathematical Chronicle*, 12:35–54, 1983.

[32] P. Erdős. My joint work with Richard Rado. In C. Whitehead, editor, *Surveys in Combinatorics*, number 123 in LMS Lecture Note, pages 53–80. Cambridge University Press, Cambridge, England, 1987.

[33] P. Erdős and R. Rado. A combinatorial theorem. *Journal of the London Mathematical Society*, 25:249–255, 1950.

[34] P. Erdős and R. Rado. Combinatorial theorems on classifications of subsets of a given set. *Proceedings of the London Mathematical Society*, 3:417–439, 1952.

[35] P. Erdős and R. Rado. A problem of ordered sets. *Journal of the London Mathematical Society*, 28:426–438, 1953.

[36] P. Erdős and R. Rado. A partition calculus in set theory. *Bulletin of the American Mathematical Society*, 62:427–489, 1956.

[37] P. Erdős and P. Turán. On some sequences of integers. *Journal of the London Mathematical Society*, 11(4):261–264, 1936.

[38] S. Fujita, C. Magnant, and K. Ozeki. Rainbow generalizations of Ramsey theory – a dynamic survey. *Theory and Applications of Graphs*, 0(1), 2014.

[39] H. Furstenberg. Ergodic behaviour of diagonal measures and a theorem of Szemerédi on arithmetic progressions. *Journal of Analysis Mathematics*. 31:204–256, 1977.

[40] H. Furstenberg and B. Weiss. Topological dynamics and combinatorial number theory. *Journal d'Analyse Mathématique*, 34:61–85, 1978.

[41] H. Furstenberg and Y. Katznelson. A density version of the Hales–Jewett theorem,. *Journal of Analysis Mathematics*, 57:64–119, 1991.

[42] W.T. Gowers. A new proof of Szemerédi's theorem. *Geometric and Functional Analysis*, 11:465–588, 2001.

[43] W.T. Gowers. Is massively collaborative mathematics possible? *Gowers's Weblog*, January 2009.

[44] R. L. Graham and B. Rothschild. A short proof of van der Waerden's theorem on arithmetic progressions. *Proceedings of the American Mathematical Society*. 42:385–386, 1974.

[45] R. L. Graham and B. Rothschild. Ramsey theory. In G–C. Rota, editor, *Studies in Combinatorics*, volume 17, pages 80–99. Math. Assoc. of Amer., Washington, D.C, 1978.

[46] R. L. Graham, B. Rothschild, and J. H. Spencer. *Ramsey Theory*. John Wiley & Sons, New York, 2nd edition, 1990.

[47] R.L. Graham. Recent developments in Ramsey theory. In Z. Cisielski and C. Olech, editors, *Proceedings of the International Congress of Mathematicians, Warsaw, 1983*, volume 2, pages 1555–1569, Amsterdam, 1984. North Holland.

[48] R.L. Graham. Some of my favorite problems in Ramsey theory. *Integers: Electronic Journal of Combinatorial Number Theory*, 7(2):A15, 2007.

[49] R.L. Graham and S. Butler. *Rudiments of Ramsey Theory*. American Mathematical Society, 2nd edition, 2015. Providence, Rhode Island.

[50] B. Green. New lower bounds for van der Waerden numbers. *Forum of Mathematics, Pi*, February 2022. 10:e18, Cambridge University Press, DOI: 10.1017/fmp.2022.12.

[51] B. Green and T. Tao. The primes contain arbitrarily long arithmetic progressions. *Annals of Mathematics*, 167:481–547, 2008.

[52] R.E. Greenwood and A.M. Gleason. Combinatorial relations and chromatic graphs. *Canadian Journal of Mathematics*, 7:1–7, 1955.

[53] A. Hajnal and J.A Larson. Partition relations. In A. Foreman and A. Kanamori, editors, *Handbook of Set Theory*, volume 1, pages 129–214. Springer, 2009.

[54] A.W. Hales and R.I. Jewett. Regularity and positional games. *Transactions of the American Mathematical Society*, 106(2):222–229, 1963.

[55] K. Hartnett. A puzzle of clever connections nears a happy end. *Quanta Magazine*, May 2017.

[56] N. Hindman and E. Tressler. The first nontrivial Hales–Jewett number is four. *Ars Combinatoria*, 113:385–390, 2014.

[57] J. Jin, L. Listiarini, T. Su, and J. Xie. A graphic novel about the creation of the proof of van der Waerden's theorem. In V. Jungić, editor, *Introduction to Ramsey Theory: 2020-2021 students' projects*. Simon Fraser University, Burnaby, BC, 2022.

[58] V. Jungić. The Never-ending happiness of Erdős's mathematics. *The Mathematical Intelligencer*, DOI: 10.1007/s00283-023-10267-5, 2023.

[59] I. Jubb. Professor George Szekeres (1911–2005), mathematician, 2005.

[60] V. Jungić. An introduction of the problem of finding the chromatic number of the plane, part I. *Crux Mathematicorum*, 45(8):390–397, October 2020.

[61] V. Jungić. An introduction of the problem of finding the chromatic number of the plane, part II. *Crux Mathematicorum*, 47(8):384–391, October 2021.

[62] V. Jungić, J. Nešetřil, and R. Radoičić. Rainbow Ramsey theory. *Integers: Electronic Journal of Combinatorial Number Theory*, 5(2):A 18, 2005.

[63] J.G. Kalbfleisch and R.G. Stanton. On the maximum number of coplanar points containing no convex n-gons. *Utilitas Math*, 47:235–245, 1995.

[64] W.M. Khelifi, C.E. Merriam, and M. Sood. The Ramsey theory podcast: No strangers at this party with David Conlon, Spotify, December 2021.

[65] A.Y. Khinchin. *Three Pearls of Number Theory*. Dover Publications, Mineola, New York, 1998.

[66] E. Klarreich. Mathematician hurls structure and disorder into century-old problem. *Quanta Magazine*, 2021.

[67] M. Kouril and J.L. Paul. The van der Waerden number $W(2,6)$ is 1132. *Experimental Mathematics*, 17(1):53–61, 2008.

[68] J. Kozik and D. Shabanov. Improved algorithms for colorings of simple hypergraphs and applications. *Journal of Combinatorial Theory, Series B*, 116:312–332, 2016.

[69] B. Kra. Ergodic methods in additive combinatorics. In A. Granville, M.B. Nathanson, and J. Solymosi, editors, *Additive Combinatorics*, volume 43 of *CRM Pro. Lecture Notes*, pages 103–144, Providence, RI, 2007. American Mathematical Society.

[70] B. Krammar, A.A. Singh, and A. Raj. The Ramsey theory podcast: No strangers at this party with Joel Spencer, Spotify, January 2022.

[71] B. Krammar, A.A. Singh, and A. Raj. The Ramsey theory podcast: No strangers at this party with Julian Sahasrabudhe, Spotify, January 2022.

[72] B. Krammer, A Raj, and A.A. Singh. The Ramsey theory podcast: No strangers at this party with Fan Chung–Graham, Spotify, January 2022.

[73] B. Landman and A. Robertson. *Ramsey Theory on the Integers*, volume 73 of *Student Math. Library*. American Math. Society, Providence, RI, 2nd edition, 2014.

[74] I.B. Leader. Ramsey Theory, 2000.

[75] W. Ledermann. Some personal reminiscences of Issai Schur. In A. Joseph, A. Melnikov, and R. Rentschler, editors, *Studies in Memory of Issai Schur*, volume 210 of *Progress in Mathematics*, pages xxxii–xxxv. Springer, Boston, Mass, 2002.

[76] Ø. Linebo. *Thin Objects: An Abstractionist Account*. Oxford University, Oxford, UK, 2018.

[77] L. Lovász. *Combinatorial Problems and Exercises*. North-Holland, Amsterdam, 1979.

[78] E. Makai Jr. Letter to Arianna Jaffer and Eugene Kim. Unpublished, November 2021.

[79] D.H. Mellor. The eponymous F.P. Ramsey. *Journal of Graph Theory*, 7(1):9–13, 1983.

[80] D.H. Mellor. Cambridge Philosophers I: F.P. Ramsey. *Philosophy*, 70(272):243–262, 1995.

[81] G. Mills. A quintessential proof of van der Waerden's theorem on arithmetic progressions. *Discrete Mathematics*. 47:117–120, 1983.

[82] J. Moreira. Monochromatic sums and products in $\mathbb{N}$. *Annals of Mathematics*, 185(3):1069–1090, 2017.

[83] L. Moser. Solution to problem E 773 [1947, 281]. *American Mathematical Monthly*, 55:99, 1948.

[84] L. Moser. *An Introduction to the Theory of Numbers*. The Trillia Lectures on Mathematics. The Trillia Group, West Lafayette, IN, 2nd edition, 2011.

[85] L. Moser and W. Moser. Solution to problem 10. *Canadian Mathematical Bulletin*, 4:187–189, 1961.

[86] O. Patashnik. $4 \times 4 \times 4$ Tic-Tac-Toe. *Mathematics Magazine*, 53(4):202–216, 1980.

[87] M. Paul. *Frank Ramsey (1903–1930) A Sister's Memoir*. Smith–Gordon in cooperation with Nishimura, 2012.

[88] G. Polya. *Mathematical Discovery: On Understanding, Learning and Teaching Problem Solving (Combined Edition)*. Wiley, Hoboken, NJ, 1981.

[89] D.H.J. Polymath. A new proof of the density Hales–Jewett theorem. *Annals of Mathematics*, 175:1283–1327, 2012.

[90] R. Rado. Studien zur kombinatorik. *Mathematische Zeitschrift volume*, 36:424–470, 1933.

[91] R. Rado. The distributive law for products of infinite series. *Quarterly Journal of Mathematics*, 11:229–242, 1940.

[92] F.P. Ramsey. On a problem of formal logic. *Proceedings of the London Mathematical Society Series 2*, 30(4):338–384, 1930.

[93] F.P. Ramsey. *The Foundations of Mathematics and Other Logical Essays*. Kegan Paul, Trench & Trubner, London, UK, 1931.

[94] H. Robbins. A remark on Stirling's formula. *American Mathematical Monthly*, 62:26–29, 1955.

[95] A. Robertson. Down the large rabbit hole. *Rocky Mountain Journal of Mathematics*, 50(1):237–253, 2020.

[96] A. Robertson. *Fundamentals of Ramsey Theory*. Discrete mathematics and its applications. Chapman & Hall, CRC Press, Boca Raton, FL, 2021.

[97] A. Robertson and D. Zeilberger. A 2-coloring of $[1, n]$ can have $(1/22)/n^2 + O(n)$ monochromatic Schur triples, but not less! *Electronic Journal of Combinatorics*, 5:R19, 1998.

[98] C.A. Rogers. Richard Rado, 28 April 1906 – 23 December 1989. *Biographical Memoirs of Fellows of the Royal Society*, 37:411–426, 1991.

[99] A.L. Rubinoff. Problem E 773. *American Mathematical Monthly*, 54:281, 1947.

[100] A. Sah. Diagonal Ramsey via effective quasirandomness. Duke Math. J. Advance Publication, 1-23 (2023). DOI: 10.1215/00127094-2022-0048.

[101] J. Sahasrabudhe. Exponential patterns in arithmetic Ramsey theory. *Acta Arithmetica*, 182:13–42, 2018.

[102] J.H. Sanders. *A generalization of Schur's thorem*. PhD thesis, Yale University, 1968.

[103] M.M. Schiffer. Issai Schur: Some personal reminiscences. In P. Duren and L. Zalcman, editors, *Menahem Max Schiffer: Selected Papers*, volume 2 of *Contemporary Mathematicians*, pages 549–555. Springer, New York, 2014.

[104] I. Schur. Uber die Kongruenz $x^m + y^m \equiv z^m \pmod{p}$. *Jber. Deutsch. Math.–Verein.*, 25:114–117, 1916.

[105] I. Schur. Über potenzreihen, die im inners des einheitskreises beschräkt sind. *Lournal für die reine und angewandte Mathematick*, 147:205–232, 1917.

[106] S. Shelah. Primitive recursive bounds for van der Waerden's numbers. *Journal of American Mathematical Society*, 1(3):683–697, 1988.

[107] M. Simonovits. Paul Erdős's influence on extremal graph theory. In R.L. Graham, J. Nešetřil, and S. Butler, editors, *The Mathematics of Paul Erdős II*, pages 245–312. Springer, New York, NY, 2nd edition, 2013.

[108] K. Singh, V. Jungić, and J.B. Mei. Life of a working Ramsey theorist: Conversation with Thomas C. Brown. *Journal of Humanistic Mathematics*, 12(1):399–407, 2022.

[109] A. Soifer. *The Mathematical Coloring Book: Mathematics of Coloring and the Colorful Life of Its Creators*. Springer, New York, 2009.

[110] A. Soifer. *The Scholar and the State: In Search of van der Waerden*. Birkhäuser, 2015.

[111] J. Spencer. Ramsey theory and Ramsey theoreticians. *Journal of. Graph Theory*, 7(1):15–23, 1983.

[112] R. S. Stevens and R. Shantaram. Computer-generated van der Waerden partitions. *Journal of Combinatorial Theory, Series A*. 32:115–118, 1982.

[113] A. Suk. On the Erdős–Szekeres convex polygon problem. *Journal of the American Mathematical Society*, 30:1047–1053, 2017.

[114] E. Szekeres. Problem E 740. *American Mathematical Monthly*, 53:462, 1946.

[115] G. Szekeres. A combinatorial problem in geometry – Reminiscences. In J. Spencer, editor, *Paul Erdős, Art of Counting, Selected Writings*, Mathematicians of our time, pages xix–xxii. The MIT Press, Cambridge, MA, 1973.

[116] G. Szekeres and L. Peters. Computer solution to the 17-point Erdős–Szekeres problem. *The ANZIAM Journal*, 48:151–164, 2006.

[117] E. Szemerédi. On sets of integers containing no four elements in arithmetic progression. *Acta Mathematica Academiae Scientiarum Hungaricae*, 20:89–104, 1969.

[118] E. Szemerédi. On sets of integers containing no *k* elements in arithmetic progression. *Acta Arithmetica*, 27:299–345, 1975.

[119] T. Tao. The ergodic and combinatorial approaches to Szemerédi's theorem. In A. Granville, M.B. Nathanson, and J. Solymosi, editors, *Additive Combinatorics*, volume 43 of *CRM Proceedings and Lecture Notes*, pages 145–193, Providence, RI, 2007. American Mathematical Society.

[120] G. Tóth and P. Valtr. The Erdős–Szekeres theorem: upper bounds and related results. In J. E. Goodman, J. Pach, and E. Welzl, editors, *Combinatorial and Computational Geometry*, volume 52 of *MSRI Publications*. Cambridge University Press, New York, NY, 2005.

[121] D. van Dalen. *Mystic, Geometer, and Intuitionist: The Life of L.E.J. Brouwer*, volume 2. Clarendon Press, Oxford, 2005.

[122] B.L. van der Waerden. Beweis einer Baudetschen vermutung. *Nieuw Archief voor Wiskunde*, 2(15):212–216, 1927.

[123] B.L. van der Waerden. How the proof of Baudet's conjecture was found. In L. Mirsky, editor, *Studies in Pure Mathematics*, pages 251–260. Academic Press, London, 1971.

[124] B.L. van der Waerden. *History of Algebra*. Springer, Berlin, Heidelberg, New York, Tokyo, 1985.

[125] J.H. van Lint. The van der Waerden conjecture: Two proofs in one year. *The Mathematical Intelligencer*, 4(5–6):72–77, 1982.

[126] D. Zagler. Newman's short proof of the prime number theorem. *American Mathematical Monthly*, 104(8):705–708, 1997.

[127] F. Zhang, editor. *The Schur Complement and Its Applications*, volume 4 of *Numerical Methods and Algorithms*. Springer, 2005.

Index

For Product Safety Concerns and Information please contact our
EU representative GPSR@taylorandfrancis.com Taylor & Francis
Verlag GmbH, Kaufingerstraße 24, 80331 München, Germany